T0301374

STIMULATING INNOVATION IN PRODUCTS AND SERVICES

STIMULATING INNOVATION IN PRODUCTS AND SERVICES
With Function Analysis and Mapping

J. JERRY KAUFMAN

ROY WOODHEAD

WILEY-INTERSCIENCE

A JOHN WILEY & SONS, INC., PUBLICATION

Published by John Wiley & Sons, Inc., Hoboken, New Jersey.
Published simultaneously in Canada.

For general information on our other products and services or for technical support, please contact our Customer Care Department within the United States at (800) 762-2974, outside the United States at (317) 572-3993 or fax (317) 572-4002.

Wiley also publishes its books in a variety of electronic formats. Some content that appears in print may not be available in electronic formats. For more information about Wiley products, visit our web site at www.wiley.com.

Library of Congress Cataloging-in-Publication Data:

Kaufman, J. Jerry.
   Stimulating innovation in products and services : with function analysis and mapping
/ by J. Jerry Kaufman, Roy Woodhead.
      p. cm.
   Includes bibliographical references of index.
   ISBN-13: 978-0-471-74060-5
   ISBN-10: 0-471-74060-8
   1. New products—Management. 2. Production planning. 3. Product management.
   4. Success in business. 5. Organizational effectiveness. 6. Corporate culture. I. Woodhead,
Roy (Roy M.) II. Title.

HF5415.153.K377 2006
658.5′038—dc22

2005050195

10 9 8 7 6 5 4 3 2

CONTENTS

FOREWORD   **xi**

PREFACE   **xiii**

ACKNOWLEDGMENTS   **xv**

1   INTRODUCTION   **1**

The Meaning of Function / 2

   Reading FAST / 4

   FAST Logic / 5

   Some Observations / 6

   What Have We Learned? / 6

Applying FAST to Hardware Products / 7

   Reading a FAST Model / 7

   Analyzing a FAST Model / 7

Some Unique Ways That a FAST Model Has Been Used / 9

How It All Began / 12

Toward an Innovation Process / 13

Who Models Functions? / 15

   Why an Interdisciplinary Team? / 16

   Team Makeup / 16

Unlocking Practical Ingenuity / 17

   When Should We Use FAST? / 17

   Fundamental Questions / 18

Distinguishing Between Problem and Opportunity / 21
Difference Between FAST Diagrams and FAST Models / 21
Validating Function Models / 23
Outline of This Book / 24

2 PROBLEM-SOLVING TECHNIQUES   26

Verb–Noun Function Technique / 29
Fuzzy Problem Technique / 31
Setting Up the Problem in the Fuzzy Problem Technique / 33
Hierarchical Technique / 38
Verb–Noun and Fuzzy Problem Techniques Within the Hierarchical
Technique / 38
Closing Remarks / 42

3 FUNCTION ANALYSIS   44

Function Analysis Syntax / 45
Active Verbs / 45
Measurable Nouns / 46
Using Two Words to Describe Functions / 47
Defining and Classifying Functions / 48
Types of Functions / 49
Extrinsic Functions / 49
Intrinsic Functions / 49
Basic Functions / 49
Secondary Functions / 50
Practical Definitions / 50
Rules Governing Basic Functions / 50
Function Identification Example / 53
Random Function Determination / 54
Levels of Abstraction / 54
Function and Component Selection / 55
Function Cost Matrix / 56
Simplifying the Process / 56
Closing Remarks / 58

4 FUNCTION ANALYSIS SYSTEM TECHNIQUE   59

Process Overview / 59
Some Misconceptions / 61
"As Is" Versus "Should Be" Models / 62
Syntax Used to Create and Read a FAST Model / 63

Reading *How–Why* and Our Intentions / 63

How–Why Versus *Why–How* Orientation / 64

Reading *When* to Consider Causality and Consequential
Functioning / 66

Key Elements of a FAST Model / 67

Scope Lines / 67

Highest-Order Function(s) / 69

Lowest-Order Function(s) / 69

Basic Function(s) / 69

Content / 70

Requirements or Specifications / 70

Dependent Functions / 70

Independent (Support) Functions / 71

Logic Path Functions / 71

Articulating Theories in FAST / 71

Variations of *How–Why* Questions / 72

Considering *And–Or* Along the Logic Path / 73

Considering *And* in the *When* Direction / 75

Considering *Or* in the *When* Direction / 75

FAST Model-Building Process: Product Example / 77

Expanding the Number of Functions / 78

Case for Using Active Verbs / 79

Purpose of Expanding Functions / 79

Avoiding Duplicate Functions / 80

Starter Kit Functions / 80

Preparations for Building a FAST Model / 80

Build *How* and Test *Why* / 80

Relationship of the Left Scope Line to the Basic Function / 81

Right Scope Line / 83

Left Scope Line / 83

What's the Problem? / 83

Defining the Problem / 84

Three Questions Before Starting the FAST Process / 85

How the Strategic Questions Are Asked in a Workshop / 86

Symbols and Notations Used in FAST Modeling / 86

Taking Exception to the FAST Rules / 87

Independent Functions Above the Logic Path, Activities Below
the Logic Path / 87

No Activities in the Major Logic Path / 88

Only Two Words Used to Describe Functions / 89

Loop-Back Modeling / 89
   Validating the Logic Flow / 90
   Exploration Drilling Model / 92
Closing Remarks / 92

5 DIMENSIONING THE FAST MODEL   **93**

Pre-event Stage / 93
FAST Dimensioning Themes / 95
Business Process and Soft Issues / 95
Sensitivity Matrix / 95
Facility Management Case Study / 96
   Determining Responsibility, Move to Action / 96
Incorporating Other Dimensions in FAST Models / 99
RACI/RASI Dimensioning / 99
FAST and Organizational Effectiveness / 101
Organizational Effectiveness Case Study / 101
   Model the Future or the Present? / 103
   Incorporating Additional Dimensions / 104
Product- and Equipment-Based FAST Models (Artifacts) / 104
   Sensitivity Matrix in Product (Artifact) Analysis / 104
Staple Remover Case Study Using FAST With the Sensitivity
Matrix / 106
   Determining Component Function–Cost Details / 106
   Proposed Solution / 108
Pipeline Case Study Using the Sensitivity Matrix / 108
Other Case-Specific Dimensions / 109
Budgeting Operating Expenses and the Sensitivity Matrix / 109
Clustering Functions / 111
Example Using Clustering / 111
Closing Remarks / 114

6 ATTRIBUTES AND THE FAST MODEL   **115**

Defining an Attribute's Range of Acceptance / 118
Ranking Attributes / 120
Incorporating Attributes Into a FAST model / 120
Linking Issues of Concern to a FAST model / 121
Construction Management Case Study / 128
Influence of Attributes and Incentives on FAST Modeling / 138
Software Acquisition Case Study / 138
Validity of a FAST Model / 140

Pre-event's Role in FAST Modeling / 143
Areas Defined by a Scope Line / 144
Resolving the Incentive Issue / 144
Determining the Incentive Earned Points Score / 147
Closing Remarks / 149

7 ENABLING INNOVATION   **151**

Analyzing FAST Models / 151
Distinguishing Outcomes and Ideas / 153
Starting to Generate Ideas / 154
Handling Negative Functions / 155
Examples of Negative Functions / 156
TRIZ and Negative Functions: Path to Creativity / 158
Defining Problems: Prerequisite to Seeking Solutions / 160
Problem Set Matrix / 160
Identifying Critical Innovation Points / 162
Realizing Innovation Through FAST Models / 162
Toward Innovation That Makes a Difference / 163
Importance of the Pre-event Phase / 163
XYZ-3 Case Study / 164
Defining XYZ-3's Problems / 164
Setting Project Goals / 165
Selecting Attributes / 166
Selecting Random Functions / 166
Constructing the FAST Model / 167
Selecting Functions to Be Brainstormed / 169
Using FAST for Brainstorming / 169
Concluding the XYZ-3 Value Study / 175
Closing Remarks / 175

8 FROM COMPETENCY TO CAPABILITY   **176**

Moving Toward Know-How and FAST models / 177
Beyond Intuition / 177
Discovering New Knowledge / 180
Management of Functionality / 185
Using FAST Modeling to Improve the Supply Chain / 186
Using FAST Modeling to Enable Shared Understanding / 187
Managing Intangible Value to Advantage / 187
Automotive Parts Case Study / 191
How Can We Unify? / 193

Functional Enquiry / 194
Closing Remarks / 199

END NOTES   **201**

REFERENCES   **218**

APPENDIX: FREQUENTLY ASKED QUESTIONS   **222**

INDEX   **227**

FOREWORD

After participating in the first Value Engineering Class ever taught at a university in 1960, my thinking process began to change. Ed Heller taught me how to write functions using a verb and a noun during the class I attended at UCLA in Los Angeles, California. This method of thinking was a technique developed by Larry Miles when he invented value analysis. It fascinated me intellectually because it seemed to stimulate my creativity. After a couple of years of unusual creative successes while conducting Value Engineering Seminars for the Sperry Rand Corporation, my boss suggested that I write down my thinking process.

A few weeks later as I attempted to write down each step, I discovered that a cause-and-effect relationship existed between functions. When I asked how a function is performed, I discovered that my answer could be expressed as a new function, and the function I asked the question of caused my answer to come into existence. Not only that, but when I asked *why* of any new function that I formulated as my answer, I discovered my answer was always the function I started with if my thinking was correct. When a different function was formulated as the answer to the *why* question, my understanding of my project expanded immediately. As I continued this process of asking these *why–how* logic questions of every new function, I began to organize the functions into a diagram which I named a FAST diagram.

I named my system of thinking *function analysis system technique* (FAST). Various diagrams have been the outgrowth of using my method of thinking, such as FAST models, Argus charts, and operational diagrams. Many analysts still refer to these logic diagrams as FAST diagrams. I never realized that 40 years later, my innovation would have led to so many other innovations or that the term FAST would have stayed so central to value engineering.

Jerry Kaufman is a leading advocate of FAST modeling and seen as one of the best practitioners in the world. For the past 10 or 12 years he has taught FAST at the SAVE International Conferences. Those who have attended his sessions are from the United States and Canada to Australia and Africa as well as Europe and Asia. For the past 30 years he has showered me with his enthusiasm by sending me various packages loaded with his latest FAST diagram innovations and projects he has completed successfully with creative results.

Jerry's recent affiliation with Roy Woodhead, a researcher from Oxford Brookes University in England, has led to yet further innovations, and it is from this that the two have been motivated, with a large number of people nagging them, to collaborate in writing this book. They have created many innovations that merge harmoniously with FAST diagrams into a systematic method of FAST modeling.

At the heart of all innovation is the concept of function. Function analysis continues to be the major tool for advancing value engineering and value management, as they help teams to invent new things and create more value. This book will help you to increase your effectiveness and capabilities into areas that you may not have explored before. It is a pleasure to know these two fine gentlemen. They have done much to teach and advance the value management profession. Their insight and examples may provide you with the understanding you need to improve your skill in this competitive field of innovation.

They continue to influence how many of us think about functions and related issues, as they serve effectively on various boards of SAVE International. This book will be a treasured addition to those books, which have influenced us all during the past several decades.

CHARLES W. BYTHEWAY, PE, CVS, FSAVE

PREFACE

In the book we explain innovation in a way that allows it to be mastered as a result of systematic and replicable methods. We dismiss the myth that creativity is a mysterious capability available only to a few people at unpredictable moments. It understands that technology is the application of practical intelligence, so the way to improve how we innovate is to use methods that help a group of people to become a synchronized and creative team. It's a book about how to manage the practical ingenuity that resides within an organization and how it can be used more efficiently and effectively to create value.

A path to systematic innovation is an essential business requirement for organizations trying to survive in modern economies. Organizations have to compete against low-cost economies and try to win consumers that demand continually higher levels of value for the money they spend. Under such hostile business conditions, those organizations that can innovate and adapt will be able to survive and prosper. Those organizations that cannot innovate systematically end up trying to sell yesterday's newspapers and will be nudged out of the marketplace by Adam Smith's invisible hand and Schumpeter's creative destruction. Our book takes readers on a journey from basic to advanced concepts that will help them survive and prosper in today's highly competitive environments.

This book is based on our experience helping major multibillion-dollar projects in the oil and gas industry, aerospace industry, health care industry, and small manufacturing firms trying to improve profitability. In the public sector we supported projects to give better returns on tax dollars. We work with leading-edge scientists in research and development programs. We explain how a distinction between solution and function as well as problem and opportunity open the door to a way of understanding how to innovate that can be achieved systematically. The key to this success is in the way functionality is made explicit and auditable.

Our real-world observations of processes are translated into abstract concepts that communicate relevance and yet do not tie us into any particular solution. We ask "What is the function performed?" in such a way that we can tease out and make explicit the functionality needed for any system, product, or project. Once we have a clear understanding of what an elegant solution should achieve, we can look at what we have and know what needs to be improved and how such improvement would yield additional value. The power of this approach enables sessions such as brainstorming or synectics to be empowered because the scientists, engineers, and managers working for the same clients understand, in a combined and practical way, why a current offering is not delivering, or will not deliver, the results that the marketplace expects. It's a book that will change the way that people inside organizations think about the act of innovation.

Purpose in Writing This Book

Many multinational corporations have for many years regarded Jerry Kaufman as an innovation catalyst. Dr. Woodhead was mentored by Mr. Kaufman in the 1990s and as a Senior Lecturer in Technology Management at Oxford Brookes University in Oxford, England, has helped to promote the way in which functions are seen as central to innovation. Supported by a history of successful commissions and comments from workshop participants and clients, we have been encouraged to compile many case studies in a book and explain how others can get the results we have achieved. This book is intended to share what we have developed and to start a quiet revolution that will see a proliferation of innovations, as the embedded practical ingenuity 'trapped' inside so many organizations is managed more effectively as an asset in its own right. It is not a book of hollow slogans and bland opinions. It is a book that explains the underlying theories that make the numerous techniques it discusses both accessible and learnable. In short, this is a book that will make a difference because it seeks to widen the ability to innovate with a combination of practical explanations of techniques and a rich exploration of the underlying principles.

Intended Audience

The primary audience for the book are those people whose business success is very dependent on innovation and who want to get better at it. This would include engineers, scientists, and managers who work inside organizations. The book will also appeal to management consultants for their use as a basis for offering services to companies.

The book has also been written in a way that will make it useful for postgraduate students. MBA students studying modes of innovating will find the book very relevant and the end-note references helpful in their studies. In summary, the book is relevant for any individual, team, or organization that needs to improve the ability to innovate.

The book is all about making thinking explicit and generating a list of feasible alternatives before committing to action. It's about using intelligence to create options by understanding solutions that gain their value from the way they enable functions to be performed. But how do you set about achieving those ambitions? We address these outcomes and many more in seeking to help make you a team innovator.

ROY WOODHEAD
J. JERRY KAUFMAN

Oxford, UK
Spring, TX
July 2005

ACKNOWLEDGMENTS

What does an academic from the UK have in common with a practitioner from the United States: the respect and admiration for function mapping and its effects on improving the innovative talents of the users of this elegant process. For this we are ever grateful to Charles W. Bytheway, the inventor of the function analysis system technique (FAST). Charles is our guiding light and counselor as he encourages us to "play" with his creation and continue to discover new applications and outcomes for his process.

We would be remiss not to recognize and extend belated thanks to our mentors, Carlos Fallon and Lawrence Miles, who taught us the true meaning of value and that to understand the dynamics of value requires an intimate knowledge and appreciation of function analysis. We very much welcome the contributions of our associates, Dan Seni and Hank Ball, for encouraging us to move into unknown areas and explore for the sake of exploration. A toast to our wives: to Harriet Kaufman, for allowing me to make this book a higher priority than attending to the problems and decisions of selling our house and building a new home, and to Helen Woodhead, for putting up with us constantly phoning and sharing ideas in the early morning hours across the Atlantic Ocean.

Our very special thanks to William F. Christopher, Publication Services, for unselfishly giving his time and talents: editing, suggesting, criticizing (when it was deserved), encouraging, and advising us, making this project one in which we take tremendous pride.

J.J.K.

CHAPTER 1

INTRODUCTION

This book is about the underlying principles that enable systematic approaches to innovation. It will help readers become better innovators by learning and understanding the principles of the *function analysis system technique* (FAST). We want to encourage readers to experiment with function analysis in new and challenging situations. Therefore, the focus is on uncovering the underlying principles as well as the procedures. Each chapter covers one or more key principles and the logic behind them. To help the reader to apply these principles, a variety of practical examples and case studies are included that explain and demonstrate the steps involved. We begin by providing a brief outline and history of the topic.

Many books have been written about innovation and techniques promising to improve the ability to innovate. What makes this book different, and how will you benefit? The book will guide you through many innovative, value-adding applications. In product analysis you will learn how to reduce development and product cost, when and how to extend product model offerings, and when to consolidate the number of models while protecting or improving the customer's value perception. You will also gain knowledge of what features to incorporate into a standard product and which features should be offered as product options. The book will also help you focus on incorporating high-valued functions in new product development ventures; to gain the competitive edge that drives the need for new, advanced products.

Applying the principles contained in this book to major projects and programs in fields such as oil and gas exploration and chemicals and their by-products will result in significant reductions in capital expenditures (capex) and operating expenses (opex) while improving process time, net present value, internal rate of return, throughput, and other projected outcomes. The processes described in the chapters that follow will

Stimulating Innovation in Products and Services: With Function Analysis and Mapping,
by J. Jerry Kaufman and Roy Woodhead
Copyright © 2006 John Wiley & Sons, Inc.

help you plan projects, develop technical and business processes and procedures, and modify organizations for leaner, more effective performance.

The technique discussed throughout the book will also help task teams innovate by changing the way we normally think about solving problems and capturing opportunities. That technique is a modeling process called the *function analysis system technique*, or as it is known and referred to by practitioners of the process, FAST. At the heart of the FAST modeling process rest two powerful questions that if asked strategically in researching for information will open the door to a wealth of knowledge waiting to be freed, hidden under levels of assumptions and misinformation. Those two questions are *how*? and *why*? The key to problem solving is knowing how to ask questions, what questions to ask, when, in what sequence, and how to interpret the answers. The FAST modeling process responds to each of these key questions.

Another important part of the FAST process described in this book is to think and speak in terms of functions. Once this process is learned, thinking and speaking functionally will allow team members to communicate with anyone, regardless of their technical or professional background. Is thinking and speaking functionally difficult to learn? No; it is no more difficult then learning to solve crossword puzzles; and like crossword puzzles, proficiency improves with practice.

THE MEANING OF FUNCTION

Function is a common word used in everyday language by many who do not appreciate the depth of its meaning. A passenger expressing concern about the erratic behavior of a bus is overheard to say: *"I don't think this bus is functioning properly."* A manager berating an employee for some infraction of the rules begins: *"Exactly what is your function in this organization, Mr. Smith?"* One friend questions another by asking, *"When do you think this function will end?* Are these examples of proper uses of the word *function*? If so, what is the function of a bus? Do you think Mr. Smith will be able to answer the question that he was asked? And can the friends explain functionally where they are or why they are there?

Have you learned anything yet about the process? Probably not, so let's start with a simple FAST model and see what we can make of it. Figure 1.1 is a FAST model describing how to make a cup of tea. Since an American created the model for a group of Englishmen as an example to demonstrate that *anything* can be modeled functionally, the model was inspected critically to ensure that the FAST model properly represented an Englishman's version of the art of making a good cup of tea. To read and appreciate the FAST model shown, we took a page from the way that manufacturers instruct buyers on how to set up and use their equipment. In addition to the principles and applications of function analysis and FAST and the direct link to innovation that is the main theme of this book, presented below is a quick reference guide to reading a FAST model.

First, note the arrows labeled *how*, *why*, and *when* in Figure 1.1 and the direction in which the arrows are pointing. The arrows tell you to what you are addressing your questions. Although we can start asking questions anyplace on the model, let's begin on the left side and question *how*. We explain later why that direction was selected. Notice that there is one block on the right side of the left dashed line. That block is a function. In fact, almost all the blocks on this model are functions. Therefore, we are going to use the description of one function to question another. The function questioned has the answer to the question asked. Ready to start?

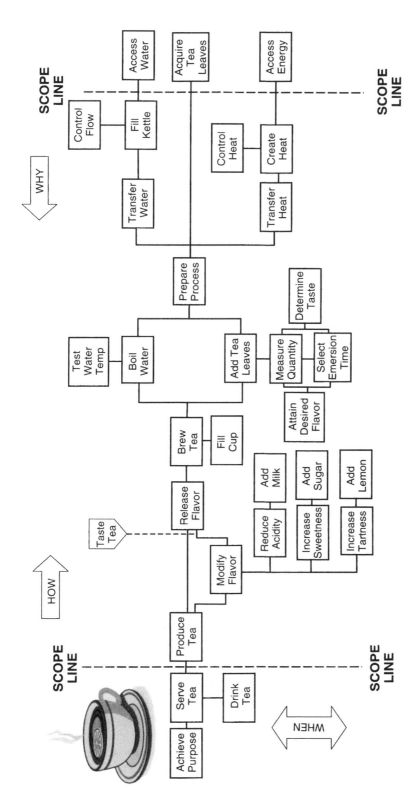

FIGURE 1.1 Brewing Tea

3

• How do you *Produce Tea?*

Produce Tea is our outcome, our goal. If we were developing a process or procedure describing how to achieve our goal, we would answer the question by jumping to the beginning and start building the procedure saying something like, "First, we get a teapot and fill it with water. Then. ..." However, Figure 1.1 is not a process or a procedure. It is a FAST model. The difference, as you will learn later in the book, is that a FAST model tells you what you must do; a *process model* tells you how to do it.

The answers to the arrowed questions asked must answer that question specifically. Answering the question "How do you *Produce Tea?*" with "First we ..." is not the specific answer to the question asked. Let's explain briefly how to read a FAST. The process is not complex, it is simply new to you. Therefore, we will move slowly until you become more comfortable with the process. In the following chapters we unfold an explanation that allows you easily to master the challenge not only of learning how to read a FAST model but also how to analyze and create one.

Reading FAST

Remember that in this introductory chapter we are merely trying to give you a flavor of FAST. This section is intended to set down some learning points that will help you realize that the FAST model helps multidisciplinary teams to share their common theory of how something works. In FAST modeling terms, the correct answer to questions such as "How do you *Produce Tea?*" can be found by looking at the functions to the right of the function questioned (i.e., to the right of the function *Produce Tea*). The *how* arrow will remind you of the left-to-right orientation. Note that we have two answers to the question, connected by a symbol, the *or* symbol, identified by two lines exiting the function asked. The two lines are optional paths and are read "How do you *Produce Tea?*" "By directly *Releasing Flavor* 'or' *Modifying Flavor*." This is "function-speak" for *"Do you want your tea straight, or flavored?"*

If you answered "flavored," the *when* arrow guides you to the response. 'When you *Modify Flavor*, you can do so by:

• *Reducing Acidity*. How? By *Adding Milk*
• Or, by *Increasing Sweetness*. How? By *Adding Sugar*
• Or, by *Increasing Tartness*. How? By *Adding Lemon*
• Or, by a combination of the three options

At this point one of our English friends said, "Ha, got you. One never combines milk and lemon in the same cup of tea." Our response was: "Maybe neither you nor we would think of that combination, but somewhere, in some faraway land, maybe there is someone who likes his tea that way. Why deprive him of the option?"

Can we question the FAST logic? Yes, that's the purpose. To explore, question, create alternative functions and generate innovative ideas to achieve the project study's objective. Notice that symbol labeled "Taste Tea"? That's an event note which advises, *"Taste the tea prior to modifying the flavor."* '

Continuing, let's ask how we *Release Flavor*. The answer is found to the right of the function questioned, *Brew Tea*. *When* we brew tea, we *Fill Cup*. But *how* do we *Brew Tea*? By *Boiling Water* 'and' *Adding Tea Leaves*. Note the *and* branch. Unlike the *or* branch, where you are given a choice of paths to follow, the *and* branch says that you must both *Boil Water 'and' Add Tea Leaves. Why*? (See the *why* arrow.) So that you can *Brew Tea*.

Let's continue along the main *how–why* path and come back to the *when* functions a bit later.

- How do we *Boil Water* and *Add Tea Leaves*? By *Preparing Process*.
- How do we *Prepare Process*? By *Transferring Water 'and' Acquiring Tea Leaves 'and' Transferring Heat*.

We can now ask the *how* question of *Transfer Heat* and *Transfer Water*, following their paths to the three functions past the right dashed line. The dashed lines in Figure 1.1, called *scope lines*, define the study boundaries. The functions to the right of the right scope line are input functions and identify the abstraction level where we wish to start the analysis. Functions to the left of the left scope line are the reason that you are, in this case, *Producing Tea*.

Now that we know *how* to *Produce Tea*, let's find out *why*. Starting with the input functions, ask:

- *Why* do we want to *Access Water*? To *Fill Kettle*.
- *Why*? To *Transfer Water* (from the tap to the pot).
- *Why* do we want to *Access Energy*? To *Create Heat*.
- *Why*? To *Transfer Heat*.

Why do we want to *Transfer Water 'and' Transfer Heat 'and' Acquire Tea Leaves*? To *Prepare Process*.

I'm sure you noticed that we passed two *when* functions: *Control Flow* and *Control Heat*. We are confident that at this point, you know how to read them. Later in the book we discuss the logic associated with the questions asked and answered.

FAST Logic

The answers to the function questions asked and answered must appeal to your sense of logic and your knowledge of making a cup of tea. When building a FAST model with a team representing different disciplines, that sense of logic must appeal to all the team members. The logic flow must also hold when following the *how* and *why* directions. If it doesn't, there is something wrong with the FAST model. Reading the FAST model in both directions is a way to test the logic flow of the model. Good FAST model builders have learned to build the model in the *how* direction and validate the model in the *why* direction. That's why we opted to start in the *how* direction.

Shall we continue validating the FAST model? *Why* do we *Prepare Process*? So we can *Boil Water 'and' Add Tea Leaves. When* you *Add Tea Leaves*, you should *Measure Quantity 'and' Select Emersion Time. How*? By *Determining Taste. Why*? So you *Attain Desired Flavor*.

You should now be able to read the FAST model with some level of understanding of the mechanics of how the three logic questions *how, why,* and *when* are used. Let's then jump to the function identified as the goal or outcome. That function, *Produce Tea*, is called the *basic function*, and all functions to the right of the basic function are *support functions*, describing in function terms the method selected to implement the basic function. What do we mean by "method selected," you ask? It means that the FAST model describes one way, among many, to *Produce Tea*. Therefore, all functions to the right of the basic function can be changed, eliminated, or combined to describe a better way of making a cup of tea, but the basic function, *Produce Tea*, cannot change. We prefer to pour hot water over a teabag in a cup but would never suggest that to an Englishman. Later in the book we show you how this insight can help you to plan whether you want radical innovation or tame innovation.

Let's now look at the functions to the left of the basic function and left scope line. We do this by continuing our *why* questions. *Why* do we want to *Produce Tea?* So that we can *Serve Tea*. *Why* do we want to *Serve Tea?* To achieve some higher purpose. That purpose could be a social gathering or some other event. If we treated the social gathering as the basic function, *Produce Tea* would be a supporting function to that event. This would allow the FAST model builder to consider a soft or "adult" drink in place of the tea. Much more will be said about functions and their important role in stimulating innovation as you read the book.

Getting back to our English friends, after a detailed examination of the FAST model, one particular fellow threw back his shoulders, raised his head, looked down his nose, sniffed, and said, "The model is invalid." "How so?" we asked. To which he replied, "You forgot to preheat the teapot."

Some Observations

Teaching FAST modeling usually starts with a hardware or artifact example so that the student can visualize and relate the product's components to their functions. However, using such examples tends to mentally lock in to the configuration, which, for the untrained, blocks innovative thinking. Our purpose in selecting a process example first to explain FAST is to try to stimulate your innovative talents while having fun doing it.

What Have We Learned?

FAST is an excellent way to create business and technical processes. FAST displays functionally what needs to be done and identifies dependent functions (*how*) and the reason for selecting those functions (*why*). Once the team, made up of those affected by the process, agrees to its function description, the process steps can be described sequentially and responsibilities assigned. Designing processes in this way reduces a significant amount of time iterating the published procedure to satisfy the concerns of those affected later.

A project plan can be developed in the same way. The function model justifies the steps, responsibilities, and resources required. Once approved, the dimensioned FAST model can be employed to produce a cost-effective outcome-focused project plan. Examples of such FAST model applications are described and displayed in subsequent chapters.

Another interesting observation relates to the visual effects of the FAST model. A common colloquial expression used by a person to verify that he or she was understood is to say, "Do you see what I'm saying?" Those unfamiliar with the expression may think, "How can one see speech?" FAST makes that colloquial expression literal and understandable. FAST is a far better process than trying to draw word images in the air, relying on the imagination of listeners to configure the same image picture as that of the speaker.

APPLYING FAST TO HARDWARE PRODUCTS

There is much, much more to learn about function analysis, FAST modeling, innovation, and how they fit together to form a powerful analysis process. However, before you start that journey through this book, let's use the knowledge gained in modeling a cup of tea and apply it to a tangible example. Figure 1.2 is a FAST model of a commonly used product, a mousetrap. At first glance, the FAST model should look familiar to you. The process of FAST model building applies equally well to all situations, in every market segment, regardless of the size and theme of the project.

Reading a FAST Model

Let's start with the basic function *Kill Mouse* and read the FAST model in the *how* direction.

- *How* do we *Kill Mouse*? By *Striking Mouse*.
- *How* do we *Strike Mouse*? By *Releasing Striker*.

Testing the logic to this point, we read back in the *why* direction.

- *Why* do we *Release Striker*? To *Strike Mouse*.
- *Why* do we *Strike Mouse*? To *Kill Mouse*.

By this time you should be fairly proficient in reading FAST model examples. You will find some complex FAST model examples in the book, but don't be intimidated by the size or area covered by a FAST model. All FAST models are read in the same way regardless of their complexity.

Analyzing a FAST Model

After constructing a FAST model, which you have yet to learn, step back and try to determine what the model is telling you. We have many years of experience in building FAST models with project teams and have produced over 850 to 950 FAST models. One thing we've noticed is that we have always learned something new about the project which stimulated innovation. This is directly attributable to the greater understanding that a team gains by making their thoughts explicit and structured. Let's give you a glimpse into how this augmented understanding is achieved and prepare you for what is to follow in the rest of the book.

Looking at the way the basic function is configured is a "messy" way to kill mice. We can't change the basic function if the primary objective is to kill mice, but we can look at the supporting functions to "build a better mousetrap." A predator such as a cat can do the job, as can mouse and rat poisons. There are many other creative ways to perform the basic function, but do we really want to *Kill Mice*? One would think that as long as they are out of your home, you wouldn't care if they lived or died. What if we were to make the function *Eliminate Mice* basic? Doing so would move the FAST

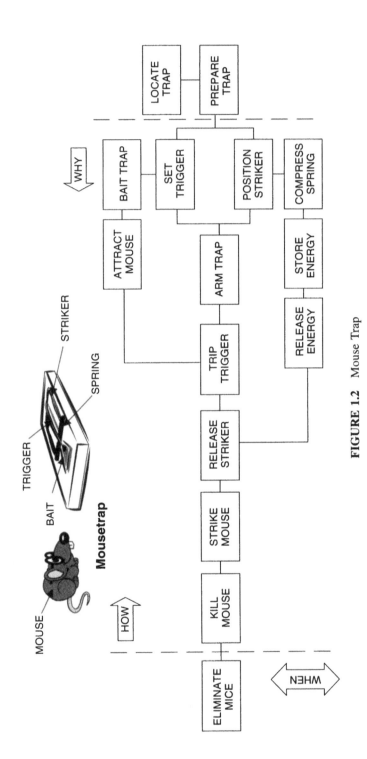

FIGURE 1.2 Mouse Trap

model to a higher level of abstraction and open many more avenues to explore for ridding ourselves of the mice problem. By making *Eliminate Mice* basic, the function *Kill Mouse* becomes a secondary function and can be altered or eliminated. After all, killing mice is but one of many ways of eliminating mice.

Let's consider a way to sterilize the mice so that they can't reproduce, or create a hostile environment by producing an annoying high-pitched ultrasonic sound that disturbs the creatures but is undetectable by humans. One participant at a FAST learning workshop suggested catapulting the mice into an unfriendly neighbor's yard. Another participant suggested a totally new approach. His idea was to capture and train the mice to eat roaches, thereby turning the liability of having mice into an asset. Another participant asked, "What is the value of the mousetrap?" The response came in the form of a question: "Who determines the trap's value, the mouse?" This may be a humorous response, but it begs a bigger question: "Who is the customer who determines the value of products and services, and what functions does the customer consider valuable?" These issues, among others, are addressed in the book.

SOME UNIQUE WAYS THAT A FAST MODEL HAS BEEN USED

There are many diverse uses for mapping and modeling functions. FAST models have been used to describe the secular and nonsecular functions of a church, a surgical procedure for replacing a knee, oil and gas exploration, reorganizing a company, functionally justifying department budgets, refining vaccine production, plotting a career path, planning a wedding, developing new products, and many other interesting and innovative project support applications—all this in addition to effective unit product cost reduction of consumer products, reducing the capital and operating expenses of multimillion-dollar international oil and gas projects while improving net present value and internal rate of return, and saving time (time to market, time to reach "first oil"), which in many cases is more important then cost reduction.

A few years ago, a survey of FAST model users was conducted to determine in what unique ways the FAST model was being employed. Following is a summary of responses collected in the survey.

1. *Communicate across technologies.* The FAST language uses active verbs and generic measurable nouns to build a FAST model. This allows communication across disciplines. FAST model building is especially effective when an interdisciplinary team is needed to solve multilevel problems.

2. *Identify problem causes.* Emerging symptoms are often an indication of a deeper-rooted problem. All too often, problem solvers treat symptoms as the cause rather than investigating what caused the symptom to emerge. Replacing a blown fuse with one having a higher capacity is an example of symptom solving. Although the power is restored, the problem remains to emerge later, possibly with more serious consequences. The *how–why* logic clearly displays the dependent relationship that links functions. If the visible symptom can be identified as a function, the dependent function logic path will help trace and find the root cause of the problem.

3. *Raise the level of abstraction.* FAST models can be constructed on any level of abstraction. When a project calls for macroanalysis of issues and decisions, a FAST model can be built to match that abstraction level. If selected issues need more detailed

microanalysis, the FAST model can isolate and address those issues of concern selectively and drill down to the information needed without disturbing the complete model.

4. *Keep project issues in focus.* The theme and construction of FAST models are based on the results of a *pre-event* or "framing" meeting with customer representatives. Once a FAST model is constructed and agreed to by the customer, the project boundaries are determined. Within those boundaries, functions that can be changed as well as those that cannot be changed are identified. Due to the dependent relationship of functions in a FAST model, proposed improvement changes to functions can be traced to determine the change effects on other functions.

5. *Separate symptoms of problems.* See response 2.

6. *Force team consensus.* The validity of a FAST model depends on achieving team consensus of all its members who built the model. This is important when an interdisciplinary team is needed and formed to address a project's problems or opportunities. Requiring team consensus assures that each team member's contributions are equal to the inputs of the other team members. It also ensures that the diverse concerns of individual team members are reflected in the completed FAST model.

7. *Temper emotions with objectivity.* Problems involving people rather than things (e.g., department reorganization) are particularly sensitive to emotional involvement and behavior of the team members involved. Managers often commission outside consultants to conduct such studies for the very reason of avoiding emotional reactions of those staff members affected. Although the consultant's recommendations may be more objective than the recommendations of those involved directly, using outside sources sets up an adversarial relationship with the affected staff. The best approach is to involve those affected by the study outcome as project team members and get the members to buy in to their recommendations. Using FAST modeling and neutral function language to reorganize departments requires team members to address the graphical representation with their territorial and authority concerns rather than confronting other team members directly. If they wish to argue an issue, the argument will be with the logic of the FAST model, not with other team members. Examples of organization case studies and their related FAST models are described in this book.

8. *Help prioritize activities.* Activities are the way that functions are implemented. However, most company budgets are estimated on how busy (activities) an organizational unit is, projected over the budget period. Understanding which functions are being addressed by the budget submitted allows management to explore different, more cost-effective activities to perform the required functions. It also allows management to prioritize those functions in terms of their contributions to the business plan, which guided management in deciding which activities to fund and which to suspend or delay.

9. *Enhance customer communications.* FAST models have been created to aid sales. These models display those functions that drive customers' perceived value. The model is then used as a visual aid in describing the product functionally to a customer and to the performance metrics of the valued functions. The FAST model is then used to compare the function performance of competitive products, which highlights the strengths (ours) and weaknesses (theirs) of the products being compared.

10. *Functionally define processes.* As described previously, the best way to determine how to do it (a process model) is first to determine what has to be done (a FAST model). There are other examples in the book proving the value of FAST in process design.

11. *Determine where to set objectives.* Constructing a high-abstraction-level FAST with senior management will clearly describe the direction of the company. Dimensioning the major functions by assigning them goals will help track performance against expectations. A more effective use of organizational FAST models is to move down to department-level FAST models. Understanding the higher-level goals will help department managers select complementary goals using the department's functions to support higher-level goals. The FAST models can then be "drilled down" to lower levels—to those organizational units supporting the goals of their reporting level. The result is the harmonious performance of individual organization units, focused on the performance of meaningful, high-level business goals.

Some characteristics of organizational FAST models are as follows:

1. *Resolve areas of responsibility.* Drawing organization charts should be the <u>last</u> things to do in restructuring an organization. The start of the process to improve the effectiveness of an organization is to understand its mission, charter, objectives, and functions needed to address those issues. Once the FAST model is complete, the functions can be clustered to show subunits by drawing unit boundaries around selected functions. Identifying which disciplines should "own" the functions could then be matched to the functions.

2. *Enhance organization effectiveness.* The boundaries around organizational units and their position on the FAST model will show which units depend on the performance of other units to meet their performance objectives. As an example, manufacturing depends on procurement to receive the correct material needed to meet the production schedule. With the dependencies displayed, joint performance goals can be crafted to encourage a cooperative operating culture.

3. *Identify input/output dependencies.* Developing dependent performance goals also addresses the issue of the format and quality of input and output performance information usable to departments that depend on such information to meet their performance commitments.

4. *Probe process vulnerabilities.* A FAST model is useful in assessing predicted process failures. By simulating the failure of a process function, the dependent relationship of functions in a FAST model will trace which functions are affected by such failures. The information can then be translated to the equipment fulfilling the malfunction, and corrective actions can be planned.

5. *Identify milestone events.* A process model is more effective in determining where to locate milestone events on a FAST model because the dependent relationship of functions displayed will show which functions need to be completed before meeting the milestone requirements. If the activities are inconsistent or in conflict with the event, the activities can be modified. An example in the manufacturing process is the location of inspection points. FAST models have helped reduce the number of inspections, and by identifying which functions will be inspected, and for what, the effectiveness of inspections has be improved and throughput time reduced.

6. *Define gate requirements.* Process gates such as an incremental funding justification are best located on the FAST model describing that process. FAST will help define who the gatekeepers are, what is required to pass through a gate, and what is needed before entering a gate. The result is a shorter process time by eliminating many loop-backs resulting from rejected funding requests.

7. *Select functions for brainstorming.* Product and process FAST models can be dimensioned to show issues of concern and which functions affect those issues directly. Instead of improving activities, brainstorming ways to determine the best way to implement those functions keeps the search for better ideas channeled to the functions concerned. This results in more innovative, responsive proposals.

8. *Outline policies and procedures.* As described previously, the consensus of what has to be done in function form provides an excellent foundation for outlining and developing policies and procedures.

9. *Help resolve technical direction.* Evaluating emerging technology by understanding and searching for new or improved ways to perform functions is a better way to match those functions to advanced products or processes. As an example, looking for new and unique ways to produce torque will result in more responses for ways to design an advanced screwdriver or drill motor than will examining a variety of competitive products.

The only thing now standing between you and learning how to achieve the results described above, and much more, is the unread chapters in this book. Your next step toward learning and applying the principles of function analysis and innovation, and benefitting greatly from this newfound knowledge, is to continue reading.

HOW IT ALL BEGAN

During World War II the U.S. government was forced to prioritize the strategic use of metals so as to ensure that the war effort was adequately supported. Manufacturers not linked directly to the war effort suddenly found themselves denied the use of critical materials such as copper, bronze, tin, zinc, ball bearings, electrical resistors, and capacitors, to name but a few. As they made products with such materials, this new constraint meant that they had no choice but to innovate or go out of business. So, through trial and error, a rapid evolution developed as other materials, such as plastics, were substituted and designs adapted quickly to ensure that the substituted materials performed in ways that did not lower product quality. After the war, manufacturers were allowed to revert back to previous designs using key metals. However, many firms did not return to old designs, as they had, in fact, found better solutions. One of the key problems facing management at that time was to figure out how they'd actually made improvements. Everyone had been so busy trying to innovate that few had actually taken notice of the innovation process they had traveled through.

At about this time, a man named Larry Miles joined General Electric (GE) in Schenectady[1] as a design engineer in the Vacuum Tube Engineering Department.[2] During his six years in that department, Miles authored 12 patents for new vacuum tubes and circuit designs. Today we would probably think of such a department as research design and development (RD&D). One day Miles asked his line manager, "Doesn't anyone at General Electric care what things cost?" This question was referred to GE's vice president of purchasing, Harry Erlicher. That question and its response had a profound effect on the theme of this book.

During World War II, Harry Erlicher observed that many forced material and component substitutions made the product not only cheaper but also better. He is credited with initiating an activity to learn how to make changes by intent rather than necessity.

Whether one person or another wins acclaim for an idea is not the point. The point is that the start of value analysis and function analysis emerged from a *learning program*. Central to the problem they faced was a way of searching for an alternative solution. This led Miles to realize that if he could not get the product, he had to get the function. This realization that a product is simply a function carrier or function enabler was at the heart of what would later become an innovation methodology called *value analysis*. In an interview, Miles was asked to summarize the emerging value methodology.

Miles described his process as "clearcut thinking" that starts by gathering knowledge. He began the process by asking; "What are we trying to do, and what are the facts surrounding the situation?"[3] Miles wanted to get more knowledge, pieced together in some logical fashion, than that "scattered around" randomly. This knowledge included function knowledge and having a clear understanding of what the customer wanted done. That knowledge included cost and cost-related functions. After this knowledge was collected, it was then put together in words that expressed clearly what needed to be done, including some things he didn't know how to do. Miles now felt that he was ready to move into creativity. He stressed that it was important to have a good knowledge base before starting to create ideas. Looking at the project, he began to develop ideas by asking: "How else might we do something? Dig into it? We all know how to do that." After collecting many ideas, Miles sorted through and selected those that offered the best way to improve the product. This was followed up by taking the necessary steps to implement the ideas. The key issues relate to knowledge, the need to understand and be creative, enquiring minds, a guiding innovation process, and a drive to achieve improvement at all times.

So what started out as a learning program to figure out how to innovate by choice led to some powerful insights into the relationship between real things, such as products and the functions embedded in their performance, from which we glean a sense of value. Miles realized that the key to innovation is functionality and knowledge and that if we can make such concepts visible and sharable, we can "manage" innovation in the same way that we can manage a schedule of activities. Miles understood that customers don't buy products but rather, the functions accessed through them. It is from such insights that his statement "All cost is for function" draws its profundity.

In the 1960s, Charles Bytheway realized that functions were logically dependent and made a significant contribution to Miles's work. Bytheway realized that as he asked what would happen if one function were eliminated, others would become unnecessary or redundant. Furthermore, he realized that one function in particular, the *basic function*, was essential for justifying all the others. He went on and developed a *how–why* logic that we explain in subsequent chapters, naming his approach the *function analysis system technique*. So this book owes a special debt to Charles Bytheway for his ingenuity and willingness to share insights. We are very proud of the fact that he also wrote the Foreword to this book.

TOWARD AN INNOVATION PROCESS

Throughout the book we discuss "things" such as products, components, and services. A "thing" does something in order to achieve an outcome. For example, the human heart beats to pump blood in order to achieve other purposes and enable other functions. The "something" that is done is the performance of a function by either a natural or an artificial solution. If we want to invent an artificial heart, we need to invent a mechanism that functions in such a way that blood is pumped around the body with an outcome comparable to that achieved by a healthy heart. *Innovation is thus concerned with the process by which we select solutions to perform functions.* Using function analysis, we can innovate systematically rather than by some mysterious and random process. Function analysis is widely used in value analysis, value engineering, and value management.[4] Function analysis enables innovation and builds on current practice drawn from a wider body of knowledge related to questions of function.[5] We encourage the reader to experiment with function analysis in new and challenging situations.[6]

Many readers may have come across the terms *function analysis* (FA) and *function analysis system technique* (FAST) in other books on value engineering (VE), value analysis (VA), and value management (VM). We refer to them collectively as *value methods*, residing within a guiding framework called *value methodology*.[7] At the heart of all value methods is the consideration of functionality accessed through the techniques of function analysis and FAST. In this book we take function analysis further, into an innovation management process that seeks to increase strategic value.

The key differences between value methods (i.e., VA, VE, and VM) are in the focus and scope to which they are applied. It is only by recognizing how something works, or should work, that innovation can follow. The "something" could be an existing part such as a car engine, a concept such as a combined hospital and university that is beyond the organization's current financial grasp, or a human resource opportunity in which a firm seeks to improve its internal capabilities by enhancing its employees' soft skills. In this book we explain a methodology that helps teams to represent visually how complex systems and products work in such a way that as a team they can use those group insights to innovate in ways that individuals cannot. This is the purpose of FAST. FAST helps to make innovation systematically available, in multidisciplinary team situations, in a way that surpasses trial-and-error invention from creative persons acting on their own. It helps managers, scientists, and engineers from different companies, industries, and universities to have their knowledge assembled in a way that achieves a practical outcome needed by organizations and required indirectly by society at large.

As far as this book is concerned, the context in which a value method is applied does not radically affect the axioms that guide the practice of building FAST models. A FAST model is simply a visual method that shows functions and their functional relationships (FAST is explained in far more detail throughout the book). The essential feature that determines whether to use VE, VA, or VM is the scope and focus used in a particular context, such as trying to improve the way an impoverished nation grows sufficient crops to feed itself or how to help a small manufacturer to reduce the cost of making a can opener. If *scope* is seen as a measure of breadth, *focus* is where the core of that scope is aimed. Scope and focus could be set at a strategic, tactical, or operational level of management thinking. We could be examining physical systems (e.g., VA) such as new products, or we could turn to social systems such as designing joint venture agreements (e.g., VM) or mixtures of social and physical systems, as in a project (e.g., VE). The important thing to note here is that the "purpose" of enabling innovation, whatever that may mean in a particular context, becomes the anchor point for the act of FAST modeling. Make no mistake about it, FAST is at the heart of the systematic innovation within the value methodologies. Let's start you on that journey by explaining how function analysis aids innovation.

Innovation is something new, different, and better. We prefer a carton of fresh milk to a carton of soured milk because one is more beneficial than the other. The concept of *beneficial* is what we mean when using the word *value* in the context of this book. Similarly, when buying a product such as a television set, we prefer today's model to older and less reliable products from 10 years ago, because of changes and differences in the way the thing works, offering us greater value. The customer, not the producer of those products or services, determines value.

Innovation is the act and realization of change and improvement. How we innovate to achieve such improvement is by implementing solutions that perform better than other solutions. A particular solution is but one way in which a function or some necessary work can be performed. Often, the functioning remains constant throughout innovative steps such as *Cook Food* as a function of a kitchen appliance, no matter how many different ways we invent to achieve it, such as a microwave oven, an electric oven, or even a gas cooker. It is simply the way functions are performed that changes. For example, consider the functioning in a car. The engine creates torque that turns the wheels. Gasoline engines, diesel engines, or electric motors, all of which are mechanical solutions that seek to perform the same function, attempt to satisfy expectations directed at the function *Transmit Torque*. Each type of mechanical solution is a different technology by which the function *Transmit Torque* can be performed, and each offers a different range of values. The technological solutions can be different but the function has a quasi-permanence that is conceptually useful to innovators.

In the design or redesign of complex objects such as an airplane or an organization, innovation is aided by making functions explicit such that we are free to swap one way of doing things for another. A function is thus a concept that characterizes what work has to be done such that the intention is clear. Because they are conceptual, we are also free to consider other solutions to the way things could be done in the real world. In this book we show you how to begin to manage the way that innovation is achieved systematically.

WHO MODELS FUNCTIONS?

Can one person develop a functional representation by working alone? The answer is "yes", and FAST models are often developed by individual practitioners to see (graphically) and critique the logic of their own thought process. However, FAST modeling is best performed by an interdisciplinary team of knowledgeable people who can share their theories and experiences of how some complex "thing" works and who also have a common interest in resolving a problem or capturing an opportunity. The common language of function, applied in a multidisciplined group, lifts the territorial boundaries that frequently separate departments. The subsequent merging of theories and ideas establishes a team with common understanding, encourages win–win cooperation, and generates insights that lead to innovation.

Team members can communicate with each other across professional vocabularies and even different languages in global projects by using the language of functions. They can gauge other team members' contributions objectively and professionally and test whether the combined theories of how something works fit with their observations and external evidence. The team will also be able to see beyond their view of the symptoms of a problem and expose the root cause. They can contemplate the impact of their decisions on other areas in the organization because of more complete oversight. Unfortunately, single-disciplined teams lack this oversight and can inadvertently create more problems than they solve because the actual functionality transcends many disciplines and departments and so requires a combined approach to innovation that is not available from a single perspective.

Why an Interdisciplinary Team?

Many years ago, an analyst facilitated a value engineering team study in the chemical industry. The team consisted of eight chemical engineers. Given the composition of the team, it shouldn't have been too surprising to learn that all the resolutions considered were chemical solutions. Unfortunately, problems are not one-dimensional. Whether the chemists knew it or not, other issues needed to be considered that were outside their knowledge domain, such as how legal, human resource management, and safety relate to each other and to the project.

After peeling away the many layers of symptoms, effects, and imposed solutions, most problems are simple to understand. With few exceptions, there are numerous solutions to problems that regularly cut across disciplines and technologies. When seeking change and innovation, we often find that there is no single "right" solution but a number of alternatives. Change and improvement are linked to many variables, such as the price of materials, investment cost, safety, reliability, and so on. To find the "best" solution requires an effective team consisting of those representatives that have a vested interest in the resolution of the problem and a clear understanding of customer expectations.

Team Makeup

Three questions govern which disciplines and associated knowledge should be represented in the team:

1. Who owns the problem or opportunity?
2. Who is responsible for the resolution of the problem or opportunity?
3. Who will be affected by the resolution?

Answering the first question, the person most affected by the problem should be in the team. This is to ensure that in defining the problem to be resolved, the correct problem, not its symptoms, will be addressed to the satisfaction of the problem owner. Symptom solving can be illustrated by a fuse that blows for no known reason and disrupts the electrical system. This is a tangible indication that a problem exists. The purpose of the fuse is to display symptoms of an electrical problem and to ensure safety. Replacing the fuse will relieve the symptom, but the problem has not been resolved until the reason why the fuse blew is identified and resolved. If the electrician does not realize that he or she owns the problem, the person will spend a lot of time replacing fuses that blow continually. In this case the electrician would be included in the team.

A toothache is another example of a symptom. Relieving the pain by taking aspirin will resolve the symptom, but the problem will remain, and increase, to cause the symptom to recur at another time. Such quick-fix masking solutions may themselves bring unforeseen side effects that lead to other problems and so make the entire situation seem far more complex than it needs to be.

Identifying who will be responsible for problem solving is easier if the problem to be resolved is defined correctly. That problem definition will "demand" which disciplines should be involved in resolving the problem. Those disciplines will also be responsible for implementing the proposed solutions. If the problem is defined as a management quandary, managers should serve on the team along with representation by those in supporting roles. If the problem is more technical in character, the technical staff should be represented as well as other staff who have insights into the problem's ramifications, such as the customer service department.

The third question, "Who will be affected by the resolution?" is important because it further ensures that the problem, not its symptoms, is being addressed. Here we would expect the voice of the customer and associated communication channels, such as the marketing department, to be represented.

UNLOCKING PRACTICAL INGENUITY

The aim of this book is to help readers become proficient in understanding methods that unlock practical ingenuity which gets us from *"How do we do something currently"* to *"How could we do it better?"* Too often, this question is not asked regularly or in a systematic innovation process. Small companies often feel that they are too busy to spend a few days innovating, so we encourage them to see this book as a way to combine work and innovation. The plain fact of the matter is that if a firm does not innovate regularly, it will be left behind as rivals offer customers superior value in a product or service. Successful companies need to improve and innovate continuously. The technology and methods of function analysis and FAST can help to minimize communication problems and hone practical know-how to achieve systematic innovation and to develop a strategic competitive advantage.

When Should We Use FAST?

Some problems are best addressed by statistical approaches, as might be expected in a market research strategy. Typical among these are *frequency* problems. In such cases we want to *count* in order to decide what to do. For example, the number of lanes on a freeway is a response to how many cars we expect will use a proposed road. Let's call this type of problem a "what to do?" problem. Methodologies from the decision sciences[8] are very appropriate for such situations, as they deal directly with observable causal accounts, relationships, and correlations between variables. But like blind men feeling different parts of an elephant, they do not give us a context-dependent insight into why such things are as they are. If after doing some surveys and plotting the trends over a five-year period we tell you that we expect the volume of traffic on a proposed freeway to reach a maximum of at 10,000 vehicles per hour, we have enabled you to decide what to do in terms of selecting a design concept for a new freeway, but you may not be thinking of other types of solutions. You are at the mercy of our assumptions because we have not described *why* this figure of 10,000 is meaningful nor how reliable it is. If we understood why the 10,000 figure was likely in terms of the purpose of many of the journeys, we might find other strategies that could reduce the need for people to travel on the freeway. We could promote alternative modes of transport, and we might find a more beneficial way to decrease congestion and improve the overall transportation system. In other words, if our generation of solutions was founded on a complete, coherent, and robust view of what happens and why (i.e., how and why it works), we will be founding solutions on an informed understanding.

So we have "what to do" and "how to do" problem types. Combining them in a single problem-solving approach will lead to better solutions by virtue of informed decision making rather than wild guessing. This is what we want to explain in this book, for the FAST model is the "how it works and why" understanding, and the way we dimension this representation gives the "what to do" insights. It is only by synthesizing the lines of inquiry within a logically valid framework that we get to understand that there are more options available to us than is often assumed. To get to this innovative capability we begin by explaining functionalism in terms of the identification of individual functions. Then we describe how to interrelate those individual functions to develop a combined logical and coherent explanation of how something works. Once we have the "how it works and why" mapped out, we will overlay the various ways in which we can identify whether functions are being performed well or not, and identify where functional dissatisfaction exists. When a solution does not perform its underlying functionality, we speak of experiencing a *malfunction*. It is the notion of malfunctions that guides our choices regarding what to do with the limited resources available to us. That is, we diagnose the health of a system just as a physician checks out your health before prescribing medication.

All we can ask of a function is—"Is this function being performed well: Yes or No?" Any further inquiry requires us to overlay a causal model and check the performance of solutions. For example, the function is not performed well because its method of performance is too expensive, its method of performance is too slow, its method of performance is too cold, or its method of performance is of poor quality. This tells us about how we perceive the performance of a function through observation of a particular solution and how we can relate that assessment to norms. But we must be wary, as we might have overlooked a causal account, such as that its method of performance is too toxic, we must use Popper's *logical falsification*[9] to push our thinking beyond the comfort zone of confirmatory evidence alone.

Fundamental Questions

Two primary and fundamental questions need to be answered: "what to do" and "how to do it." We can then address other types of fundamental questions from within a logical stability flowing from the "what to do" and "how it works and why" lines of thinking. Decisions such as those of "when" and "where" fall out of the conversations the group will have during dialogue about "what to" and "how and why to". The issues raised will be tested later so that evidence-based decision making will take priority over opinion-based decisions.

If the fundamental question were aimed at a relatively simple problem such as "What should I have for dessert?", it would not be appropriate to call a team of people together to address it. Similarly, if we were to consider radioactive contamination in the Irish Sea, we would not think about asking actors from London's West End for their advice. In other words, we need to see function analysis and the FAST model as being applied to situations in which a group decision is needed and also that an appropriate level of knowledge is brought to that group by its members.

A good rule to remember is:

- The quality of decisions cannot consistently rise above the quality of information upon which those decisions are made.

However, a corollary to the rule is:

- Considering time constraints, there is never enough time to make a no-risk decision.

The principles and application of FAST modeling add value to the decision-making process by allowing us to see where appropriate knowledge needs to be applied to improve benefits and value. Let us be clear that appropriate knowledge is of the utmost importance and that our use of the word *model* leans closer to science than to art. If we are to use the "how does it work and why" value of a FAST model to direct our thinking toward "what to do," "when to do it," and "where to do it" questions, it must be based on a valid representation of reality, or we could be led astray. In other words, if we are in the process of dealing with radioactive contamination, our view of how it works and why that is so must be correct and complete or we might be making things worse. The quality of any decision therefore depends on its relationship to an objective truth value in the lead-up to the act of deciding. The quality of a FAST model is thus assessed in terms of its proximity to a true representation of what is going on. This is central to our approach to innovation and distinguishes us from the mysterious ways of achieving creativity that rely on faith in "creative" experts. Based on our work we know that innovation can be achieved through the building and analysis of a FAST model that enables a collection of knowledgeable people to become a unified creativity team.

Let us summarize what has been said so far. This book is designed to help readers develop a command of a number of techniques that allow functionality to be understood by a multidisciplinary team in a clear and explicit way. This modeling will lead to better insights and thus better ideas suited to "complex" nontrivial problems and opportunities that require teams to address them. When our dissatisfaction is targeted on functions that do not perform as well as desired, we are addressing malfunctions. The techniques discussed in this book are based on representing how something works in a truth-laden scientific sense of correlation with evidence and validity of the explanation. It focuses on clarifying the "how and why" to achieve a "what" so that we can determine clearly "when" and "where." The act of building a FAST model is about encouraging rich dialogue that forces a team to think very clearly and to test their unspoken assumptions.

Before moving on to explain how the book is structured, let's look at an example that will be useful. Imagine that we are a new product development team and want to invent a family of office lights. Think of the strip fluorescent lights we often see strung across the ceilings of plush offices in Houston, London, or Berlin. Our task is to create a new product platform that sells thousands of units and generates significant revenue. How would we go about that? We would start by thinking about the customers and what would make them purchase the new lights. We would talk about what the lights looked like and what they were designed to do. Then we would develop a concept design and start thinking about specifications. Finally, we would select components to meet the specifications.

Now imagine that we know nothing of the above and have simply received the finished detail design and are about to start manufacturing the new lights. Let's assume that due to competition, we need to reduce the cost of manufacturing the light fixtures and light arrays and at the same time increase their performance quality so that we can sell more units and make more money. To consider possible trade-offs, we will have to start thinking about attributes that drive value choices (i.e., how we can influence customer preferences so that they will buy our product rather than our competitor's). Then we will have to look at the concept design and specifications and try to determine why the light's design is as it is before suggesting changes. If a certain component (e.g., a ballast) is specified, we might ask the question "What does it do?" to determine why it is deemed necessary. To avoid compromising some other requirement, we also need to be sure that we understand why a customer would want that particular component. In other words, we would work our way systematically from the tangible product to the designer's original intention and beyond to the causal relationships and science on which the designer's choices were made.

A function can be performed by a single solution that would yield a single outcome or range of outcomes. A different solution for the same function could yield a different outcome or a different range of outcomes. Figure 1.3 is meant to convey that by defining a function, we are able to contemplate the range of outcomes that is possible from a collection of alternative solutions. Later in the book we strengthen our discussion of the systemic nature of functionality so that we realize that the "best" solution is not necessarily selected in a one-to-one consideration of a single function but how its inclusion is synthesized at the total product level. For example, in an attempt to develop a low-cost product we might select the most expensive component to find a lower-cost way to perform the product's function. Then we wouldn't need to analyze many other components.

By starting our study off with a real product and tearing it down in a reverse engineering methodology, we have uncovered how the original designers thought about what had to be done. This allows us to make the original design intent explicit. When the designer decides that *xyz* needs to be done to *abc*, he or she is drawing upon knowledge that the person has learned. The validity of this knowledge rests outside the person and is often taught in universities. What we need is a way to tap into

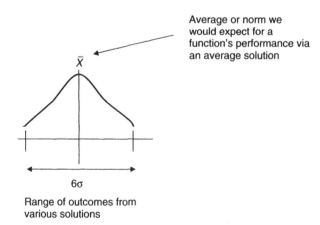

FIGURE 1.3 Normative Relationship Between Functions and Solutions That Perform Them

scientific and engineering know-how in a systematic way and in an approach that can be tested and checked for errors. Many designers will tell you that they "do" function analysis and don't need FAST models, and in many cases they may be right. However, they do not reveal their intentions explicitly in a systematic way and in a way that other disciplines can engage with and so contribute to the design theories; the team's potential is therefore reduced to following the dictates of one person.

Distinguishing Between Problem and Opportunity

We must also distinguish what we mean by problems and opportunities, to help managers give priority to areas that most need innovation and to determine which responses can be delayed as being more of a long-term planning issue. For us, a *problem* is a situation in which a function's norm is being performed unfavorably, less than or more than expected, so our problem is to find ways of aligning the function's norm with the solution's performance (see Figure 1.4). For example, following a television advertising campaign, customers expect a new car to achieve a mileage rating 40 miles per gallon but after buying it find that in use it achieves only 25 miles per gallon. That's a problem for the producer, who controlled which solutions would be implemented to perform functions, and for consumers, who are experiencing fewer benefits than they were led to believe they would get for a given sum of money. An *opportunity* is when we have a new solution that moves the norm itself. For example, we could use a new technology that shifts the norm from 40 miles per gallon to 70 miles per gallon.

Just as when a ship is sinking, keeping it afloat will be a more important problem to solve than organizing a dinner dance to improve catering revenue, managers need to prioritize how resources are to be deployed strategically. If an underresourced project is overbudget and behind schedule, the managers may wish to resolve the immediate cash flow problems before chasing tomorrow's opportunities. This is why a clear distinction between problem and opportunity is useful and in a later chapter will assist FAST models to become helpful to managers following their dimensioning and analysis.

DIFFERENCE BETWEEN FAST DIAGRAMS AND FAST MODELS

For us a *diagram* is a static representation of where things are located. A *model* is a representation of a dynamic situation in which things change. If our goal is about

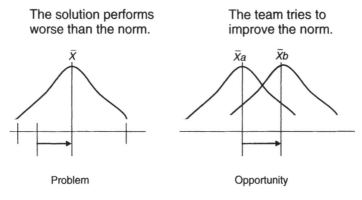

FIGURE 1.4 Understanding Problems and Opportunities

enabling innovation, we must be sure that our model portrays an accurate representation of reality. For example, if we are trying to develop a new polymer, when we suggest that *xyz* will happen as a catalyst is introduced to the feedstock, we must be sure that our inferences are reliable. This correspondence between a FAST model and what happens in reality marks a stark change from the common practice of value engineering. We do not see the FAST model as a stimulus for ideas alone; we see it as an active tool that helps us to build a rich understanding of dynamic functional relationships. In our approach to modeling we would seek confirmation that we have identified functions correctly. We also check for correlation between performance attributes to ensure that we are measuring solutions correctly. A measure of value added in the outcome is also desired. It is only when we are confident that our model is valid and adequate in an objective sense that we can start to think in a systematic way about the need for new ideas and new solutions.

The ultimate test of validity would be to build a prototype system, a pilot system, or a simulation model that functions or works mechanistically according to the logic in our function model,[10] that is, corroborate the FAST model to external data. An analogy would be the way a map represents a physical reality, just as a particular map of Oxford conveys knowledge of the location of real streets and real buildings that it reflects symbolically. If a map of Oxford bears little resemblance to the actual city layout, we would all agree that it is a poor map.

Some practitioners see internal consensus as the only way to validate good functional explanations in the form of FAST models. That a few people agree among themselves that *xyz* is the purpose being considered is an example that fails to seek out additional external knowledge and corroborating evidence. We use the test of fit from a wide perspective to allow us to glimpse the validity of our models. We challenge limitations to validity and encourage external searches for data and evidence in order to tease out and validate those assumptions that we make inadvertently. Subjectivity can be misleading, or even abused, as some team members with hidden agendas sway consensus. Rather than corresponding to objective truths, their persuasion skills promote their "subjective" opinions through coercion. The search for evidence outside the workshop is vital if we are to avoid treating shared opinions as if they were reliable facts. Innovation must always result in value being delivered as outcomes in an external strategic context, so we need both internal and external validity.

If the outcome is not recognizable by people outside the team, we do not really know if we are enabling progress and so may not actually be contributing to beneficial outcomes or value. We need to link the external strategic context, where innovation as value will be measured, to the internal decision context of the study. The internal context of the team is where the FAST model is developed, analyzed, and used to invent. Later, new solutions will be allocated to functions to see whether value can be increased. This bridging of the internal and external contexts will encourage objective arguments to steer thinking rather than personal agendas. Having said that, it is still in the organization's interest to have even a weak model-building process because if it does nothing else, it allows an opportunity to make sense of things as a group. That is, even poorly crafted consensus is better than randomness.

Our anecdotal evidence based on informal feedback suggests that those value engineering (VE) practitioners who work with a more rational approach to innovation do seem to get more reliable results than the relativists.[11] This is so because of the need to relate functions and proposed solutions to evidence and experiment so as to bridge the gap between conceptualization and realization. That is, the nonhuman world comprises functioning phenomena, such as electrons that don't negotiate, so objective data from such objective sources provide stability for our thinking, as our intentions are constrained by nature.

Nature is "the" conditioning environment we live within, and innovation is always playing with relationships of one sort or another within that environment.[12] Rationalist[13] VE practitioners often work with "gut feel" data, so an element of "nominal" social constructivism[14] and agreement is a part of their approach. We accept human knowledge as a product of data perceived by means of our five senses but believe that we can reason beyond such sensory limitations and use FAST models to help us. The notion of function is a conceptual tool that helps us to reason beyond our limited perceptions and is why some FAST modelers argue the case for the intuitive recognition of functions; they are tools to help structure our reasoning.

We want to extend the innovative capability of VE and so argue that for a value research project, subjective data must be minimized so that we can develop a rational and objective "realist"[15] view of value; we need rules that reduce the guesswork in intuitive perspectives.

Validating Function Models

FAST models and FAST diagrams are both useful to coordinate the various layers and types of design logic in complex projects to show change and consequence. The next level of awareness is how artifacts and components perform, and this is often how customers determine the utility of a particular solution, such as a keyboard or screen. Such solutions usually manipulate nature, such as getting electrons to strike a chemical coating, and when one does, the screen gives out a colored-light emission that is part of what we see when we turn a computer screen on. Underneath this line of inquiry are the designer's intentions. The designer wants to take scientific findings and put them to use. The "do x in order to achieve y" and the "if x, then y" logic of the designer is made explicit in a FAST model. A FAST diagram simply shows where "static" concepts of function are placed with respect to each other. The FAST model represents a dynamic system that is working and is alive. The FAST models help scientists to understand what engineers are trying to achieve, and helps engineers to understand why management wants certain outcomes, as well as enabling other disciplines to see how they contribute in an organizational act of collaborative innovation. A FAST model enables us to understand and manage both what is happening in one time period and what will happen in a later period through better knowledge and insight management. We can also adapt the technique of variance analysis[16] to identify the effect of a small change in the performance of one function's attribute on other functions.

The entire interest in models rather than diagrams marks a distinct change from "checklist" cultures and sometimes mandatory insistence on applying specified management techniques at key project milestones. FAST models enable managers to choose to intervene in specific ways, after diagnoses. Therefore, a rigorous approach to value methodologies with FAST models at their core can empower senior management to take charge of the systems for which they are responsible.

By using a FAST model to coordinate all the types of thinking involved in innovation, we can set about managing the entire process systematically (see Figure 1.5). We can link the organization's strategic ambitions to the external environment as well as

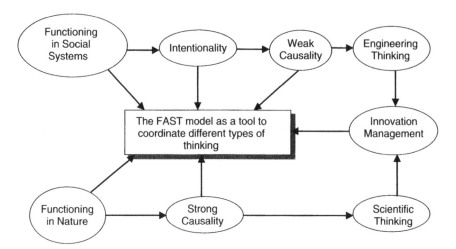

FIGURE 1.5   FAST as an Innovation Coordination Model

its tactical and operational thinking related to projects. By assessing the health of the functions in those situations, we can determine whether a particular project needs *lean thinking* to improve flow efficiency, *six sigma* to reduce variability, *business process reengineering* to rethink organizational design, and so on. As functionality is about how a system works, other management techniques can be brought into a coordinated program of innovation. Instead of following management fads, appropriate techniques are selected for a particular type of problem that our function analysis and FAST models have enabled us to recognize.

OUTLINE OF THIS BOOK

In Chapter 2 we look at how thinking functionally can help us to understand problems and opportunities better. We spend all of Chapter 3 discussing function analysis, which allows us to move toward the synthesis of a number of approaches in Chapter 4. In Chapter 3 we explain the underlying theories for the principles and techniques based around a consideration of functionality and how it can be innovated. In Chapter 4 we develop this capability so that larger and dynamic systems, such as an organization itself, can be represented as intentional and causal functions that relate and work together by way of a model. In Chapter 4 we focus on linking functions together to form explanations of how a system or product works and the function analysis system technique (FAST) procedure.

In Chapter 5 we explain how current solutions to the FAST model are linked and how such dimensioning enables insights. It is in this chapter that we learn how to identify where innovation can best be targeted. In Chapter 6 we explain the linkage from strategic problems and opportunities to the FAST model and explain how this book aids organizational decision making. In Chapter 6 we give more insight into problem and opportunity framing and how the FAST model is aligned to significant issues of concern. In Chapter 7 we move the discussion forward and explain how FAST modeling can be used to enable innovation. In the final chapter, Chapter 8, we discuss how a FAST model can be used to provide competitive advantage to an organization and the projects that it runs. The chapters are written as stand-alone units, so if you prefer to dip in and out, that will be fine. However, our intention in writing the book was to develop a structured learning process, with later chapters requiring knowledge of what was discussed in earlier chapters. Reading from Chapter 1 through to the end will help in understanding and applying the techniques more effectively.

CHAPTER 2

PROBLEM-SOLVING TECHNIQUES

Imagine a patient arriving at a doctor's office and saying something like, "I have a pain in my back and I want it to go away." The first task of the doctor is to understand the problem. In any rational approach to achieving improvement, diagnosis must always precede remedy. One of the key difficulties in problem solving is defining the problem in a way that leads to significant improvement. From a position of partial understanding we tend to react to the symptoms or outcomes of a problem rather than to the cause. In any problem-solving venture, it is critical that the correct problem be addressed and resolved. For the doctor the first line of enquiry is to find out what factors may have contributed to the patient's pained back, to discern if the cause is internal or external. Solving symptoms simply by prescribing short-term tonics could result in the emergence of the target problem in some other area, with greater consequences than those of the original problem. Problem identification techniques such as system dynamics[1] will help disclose the true nature of the problem and avoid masking or solving symptoms.

How often have you heard a manager address a newly formed team with a motivating speech something like: "Cost reduction is our biggest problem. If we can reduce our cost to produce products, our problems would be solved." There are two major faults with this statement. First, cost reduction is not a problem; it is a solution, a way to resolve a business-related problem. Second, if you agree that cost reduction is a solution, we must ask: What problems will be solved by reducing costs? The answer to that question will lead to the true problems and perhaps to more effective ways to resolve them.

Very often, good solutions are developed for the wrong problem. The results are often expensive and may not address the root cause and so fail to halt the oppressive nature of some problems. In medical terms this is like taking painkillers to mask

Stimulating Innovation in Products and Services: With Function Analysis and Mapping,
by J. Jerry Kaufman and Roy Woodhead

the effects of a degenerative disease. An example that illustrates this arose when a seemingly successful cost reduction program focused on the standard cost (and price) of a line of thermocouples. It did nothing to reverse the downward trend in lost sales. The cost reduction assumption addressed the wrong problem. A value management (VM) team modeled the functions in a wider system and later found that the problem was based in the response time for filling orders, not in the price of the product. Sales improved after the VM team's recommendations were implemented and the order-entry process was changed to fill orders in two to three days rather than the existing one to three weeks.

Another example comes from the city of New York, where VM was employed to reduce the eight-plus years it took to refurbish and modernize high schools. The problem became apparent when a newspaper editorial questioned why it took over eight years to refurbish high school facilities in New York when it only took Donald Trump about two years to build the Trump Towers. Before beginning the VM process, the team discussed the best way to come up with time-saving ideas. When the problem was better defined it was realized that the eight years was the symptom of the problem, not a cause. To remove the symptom, the VM team had to search within the system that caused it by making its functioning explicit and then testing the efficiency of the solutions used to perform those functions. The problem was determined to be that no single person or agency took responsibility for managing the complete renovation process. The process flow diagrams depicted what had to happen but were not connected to how the goal was to be achieved. This shows how process models often assume that the way things work is clear, but the reality proves otherwise. The basic problem was in the organizational structure. There was no single point of accountability to coordinate the 27 agencies involved in the high school renovation and modernization projects. The realization had been made that some "assumed" functionality was missing; an organizational realignment, along with procedural changes proposed by the VM team, was implemented to reverse the loss of value. The project time from initiation to completion was reduced by more than 50 percent.

One of the reasons that problem solving is difficult is that we tend to address that which is bothering us, the symptoms of the problem, and don't devote enough time to uncovering the true culprit, the root cause. As John Dewey said, "A problem well defined is a problem half solved." In our example of the patient at the doctor's, this search for understanding takes up most of the doctor's time. Once the doctor understands the problem and identifies ways to deal with it, treatment becomes routine. The insights come from understanding why what is supposed to happen doesn't. How often we admire those who can get right to the heart of a problem. They seem to have the ability to "see" the structure that yields the problem, to identify where in the structure the problem could best be resolved, and communicate this information in a manner that provides us with a common starting point for solutions.

The word *problem* as it is used to describe problem-solving methods more generally is not restricted to negative connotations. The ability to capture an opportunity will also fit the problem definition because to close the gap between where we are and where we want to be requires resources we might not have. However, we wish to be more precise in our thinking. We have already distinguished between problems and opportunities in terms of how a function is performed, so now we have the ability to see through unnecessary ambiguity. A problem relates to the relationships between actual performance and the norm we expected. Opportunities relate to ways in which we can move the norms themselves.

Problems and opportunities come in a large variety of sizes and shapes. There is no formula that will crank out a solution automatically. There are, however, techniques to assist in structuring and testing our theories of what is going on. FAST modeling provides a way of gaining insights into problem situations and defining them in ways that encourage the development of solutions: that is, ways in which we can represent our theories and test them in order to glean insights that lead to breakthrough innovation. Some of the techniques presented are more appropriate to certain types of problems than others, as discussed briefly in Chapter 1. During the strategic problem framing process (known as the *pre-event* in value engineering terminology), possible solutions will come to mind. It may be helpful to write these down for later reference, but their acceptability at this embryonic stage should not be explored or judged, as we have yet to develop a more complete view of what is going on.

The techniques we present are platforms that can be built upon. After becoming familiar with them, we can move forward and consider logical relationships that will permit the design of a framework for a specific problem. The problem identification techniques covered in this chapter are:

- *Verb–noun technique*: used to name an individual function
- *Fuzzy problem technique*: used to structure our understanding of the context for the study
- *Hierarchical technique*: used to say what a thing is supposed to do

These techniques are selected for discussion because they form the basis for FAST modeling. Each of these problem-solving techniques has had a major influence on the way that FAST modeling has evolved over the last 40 to 50 years. An appreciation of these techniques will enable you to better understand the principles and effective application of FAST models in resolving a variety of problem situations.

The act of modeling the problem should be structured until you are satisfied that the real problem is identified and this is verified by other checks. For example, a patient should receive treatment only if the doctor is clear:

- As to what is malfunctioning
- As to the way in which malfunctions are measured, recorded, and verified (i.e., accounted for)
- That a logical coherence exists among the initial diagnosis, the detailed investigation, and the treatment selected

Given the same problem at a different point in time or in the hands of a different problem solver, the end results can be quite different, just as one doctor would prescribe one drug and another doctor a different one. Both doctors seek the same outcome but by means of different solutions to the same malfunction. However, all problem identification techniques have one common characteristic. They approach the problem objectively and ensure that the model is an attempt at a truthful representation of what is going on.

VERB–NOUN FUNCTION TECHNIQUE

To achieve a purpose, we have to do something. This something to be done is what we mean by *function*. It is a generic statement of an abstract process that is essential to the working of the system within which it resides. To help us to think clearly, this is assisted by a naming protocol that uses a verb–noun grouping as a description of a function. In all problem-solving techniques we are trying to change a condition (e.g., an outcome) by means of a new solution that is better suited to the task. The important thing to note here is that we are trying to change a condition. The way that something is today is its condition, and we seek to innovate in order to improve on that condition. We want to intervene in a system and change things so that we get to the outcome we desire. If we describe in detail, or dialogue, what we are trying to accomplish, we tend to describe a solution and miss the opportunity to develop informed insights and engage in divergent thinking about other alternatives.

Most business problems offered to value engineering (VE) are expressed as cost reduction problems. Senior managers, who do not normally define a problem, inadvertently mandate a solution to a problem which often has not been defined adequately. In so doing, the managers unconsciously limit the search for options. The value study team would focus their minds on ways to reduce cost, which may not solve the real problem. What managers should ask is: What problem or condition will be corrected if the cost reduction goal is achieved? If the aim is to improve a product's competitive position while protecting margins, or improving sales, or a similar result, the problem resides in the underlying conditions. Cost reduction would be one method to achieve the objectives and resolve the problem, but cost reduction may not be the most effective solution. The basis of good selection is to start with numerous options to choose from; limiting the number of options, even if they appear to be good, runs contrary to this truism.

When trying to describe problems that affect us, we often become locked into a course of action without realizing it, because of our own bias. Conversely, the more abstractly we can define a function we are trying to accomplish, the more opportunities we will have for divergent thinking. This clarity of *intended action needed to change something* can be achieved by describing what is to be accomplished with a verb and a noun. In this we mean that a function is something that represents an action performed on a target to achieve an outcome that we desire. In this technique, the verb answers the question, "What action is to be done?" The verb defines the action required. The noun answers the question, to what is the action being applied? The noun tells us what object is being acted on.

Identifying a function with a verb–noun descriptor is not as simple as it might appear. As an example, what is the function of a cigarette lighter? An obvious first answer is to *Light Cigarette*. Does it also provide a function for lighting cigars, pipes, and bonfires? Upon reflection, the function might be restated as *Produce Flame* so that its use enables many purposes, each of which is context dependent. If the function is *Produce Flame*, what is the function of a flameless electric cigarette lighter in a car? Further reflection might disclose that a more appropriate description of its function is to *Produce Heat*. In such a case we could now see flame as but one solution among others, to achieve the "combust" objective.

A cigarette lighter can be described as a mechanism that provides all of the functions listed above. Since the problem, or purpose to be achieved, has not been stated, there is no basis to select the proper function. If the problem were to develop a new artifact for igniting various tobacco products (cigarette, pipes, cigars), we could say that the function is to produce heat. In a subsequent brainstorming session, the problem might be to determine how we can *Produce Heat*. This would provide a broader range of possible solutions than *Produce Flame*. The constraints of how much heat, for what duration, how portable, at what cost, and other considerations would be questioned after the ideas were generated to obtain relevant solutions for *Produce Heat*.

Given a different problem for the cigarette lighter, we might identify a different function. If we were in the business of selling lighter fluid (fuel) and wanted to market a new cigarette lighter, our end function would probably be to *Ignite Fuel*. The relationships among purpose, usefulness, and function are inextricably linked and are often governed by organizational intentions such as market leadership. Identifying the function in the broadest possible terms frees us from preconceived solutions and provides the greatest potential for divergent thinking and developing creative alternatives.

A function should be identified in terms of what is to be accomplished by a solution (i.e., how a solution contributes a vital necessity in order to achieve a purpose). A function is not a solution (i.e., its method of performance). A function makes a logically valuable contribution to the system in which it resides, whereas a process need not. For example, consider the fastening of a simple nameplate to identify a piece of equipment. Rather than specifying the function *Screw Nameplate* (i.e., a prescriptive method), the function would be better specified as *Label Equipment*. Attaching a nameplate with screws is only one of the many ways of identifying equipment. Nameplates can be riveted, welded, hung, cemented, or wired in place. The information required could even be etched, stenciled, or stamped on the equipment, thus eliminating the need for a nameplate. This must be seen as a process of seeing the design of intended actions in a very clear and precise form, which opens opportunities for innovation.

Using a verb–noun description, the identification of what is to be accomplished within the context of a single function permits the classifying of all ideas into at least two categories: those that can satisfy the function *Produce Heat* and those that fail to satisfy the function and cannot *Produce Heat*. Here is our first glimpse of the realization that functions are normative.

It is worth noting that all the ideas that satisfy the function constitute a *class* of information. The ideas (i.e., potential solutions) all have a common property in that they *Produce Heat*. As "things" have other properties, we can select other components in terms of the impact on other functions, thus making our view of what will happen more encompassing. How the function is identified determines the scope or range of solutions that can be considered. In the cigarette lighter example there are two stated levels of problems, the functions *Produce Heat* and *Produce Flame*. The former will contain a larger class of ideas than the latter because it is less of a preconceived solution. However, each is an acceptable function for the respective problem (i.e., purpose) being addressed.

Some verbs and nouns that may be helpful in identifying a function that is needed to achieve a purpose are listed in Table 2.1. Any verb may be combined with any noun in the list to describe a desired function. The acid test is to ask: "Does it reflect what a thing actually does with respect to its valuable contribution to the system in which it resides?" A function is the concept described by the verb–noun, and this is why two teams may name the same function with different words but know they are talking

TABLE 2.1 Selected Verbs and Nouns

Verbs		Nouns	
Add	Isolate	Air	Motion
Analyze	Locate	Area	Noise
Arrange	Maintain	Assembly	Opening
Attach	Measure	Atmosphere	Pressure
Attract	Minimize	Cold	Protection
Create	Obtain	Color	Resistance
Collect	Position	Comfort	Rotation
Combine	Prevent	Communication	Shape
Confirm	Protect	Component	Size
Contain	Recommend	Current	Solid
Conduct	Record	Distance	Sound
Control	Reduce	Enclosure	Space
Convert	Remove	Energy	Stress
Cool	Resist	Environment	Temperature
Destroy	Restrict	Expansion	Texture
Develop	Retain	Fluid	Time
Display	Reverse	Force	Torque
Distribute	Rotate	Frequency	Vehicle
Eliminate	Select	Friction	Vibration
Evaluate	Sense	Gas	Volume
Expand	Separate	Heat	Voltage
Extend	Start	Humidity	Waste
Freeze	Store	Indication	Water
Harden	Support	Information	Wear
Heat	Test	Length	Weight
Implement	Transmit	Material	
Increase	Transport	Mixture	
Insulate	Use	Mobility	
Invert	Verify	Moisture	

about the same underlying concept: for example, *Distribute Products* and *Dispatch Goods*. We must be wary of thinking that just because we can string a verb and noun together, that would somehow designate it automatically as a function; the verb–noun is simply a signpost to the underlying concept of function.

FUZZY PROBLEM TECHNIQUE

Without analysis of functions and functioning, many approaches to creative problem solving lack an investigation of how something works. As such, the way that things work is too often assumed to be complex, and innovation itself is seen as a result of luck rather than as a structured process. In such approaches teams are encouraged to think "outside the box" in a variety of methods thought to yield a new idea. This is good in new situations, but most companies work with established technologies and legal and economic frameworks, so the new situation simply does not exist. In our approach we see the world as comprising actions and reactions that bind science, engineering, philosophy, politics, commerce, and technology management. If we want to drill for oil in Alaska, we can clearly envisage that we will have to deal with very cold temperatures, the politics and ethics of environmentalism, commercial viability, and the availability of trusted technologies capable of getting us to the outcome we intend in the way we desire.

That is, we must think of means and ends as a complex situation that can be represented in a FAST model that enables us to have concurrent overviews of the macro, mesa, and micro variables at play. Such a capability is a basic requirement for socially responsible decision making, as in the case of projects that seek to access oil within areas of natural beauty. We have to think about the nuts and bolts at the same time that we think about the big picture.

In the *fuzzy problem technique* approach, we do not focus on the search for malfunctions. Our aim is to show how we get to a sense of purpose that steers the development of the FAST model within the typical time constraints placed on the common practice of value methodologies such as VA, VE, and VM. A strategic problem framing workshop establishes a view of which problems or opportunities management want to address. Such a workshop enables senior management commitment and psychological ownership of the specific purpose they wish to see form the focus of the value study. As strategic problems become more complex and poorly understood, it becomes more difficult to identify the functions we are attempting to satisfy. This is usually because there is more than one system at play (e.g., political systems, economic systems, social systems, technological systems, and environmental systems). The reality is that systems are not always predictable, and senior managers are not always available to demonstrate their commitment, direction, and leadership. We have to accept that ambiguity is present in our preliminary work, which makes the testing of our theories even more important. Often, we cannot even find the causes of malfunctions, as they are hidden within a confusing mess of facts, opinions, assumptions, and emotions. Being confronted with the accusation "You just don't understand the problem" is often due to a lack of agreement on the cause of the malfunction. Until there is concurrence on the problem (i.e., the malfunction is identified), there can be no agreement on the solution. We must be clear on the purpose and the functional relationships before we contemplate possible solutions.

The fuzzy problem technique offers a way to clarify those problems and opportunities that we just can't seem to get a handle on and then determine the range or scope of possibilities that are available for consideration. Most of us have avoided attempting to solve a problem at one time or another because the problem was not clear enough in our mind to attempt a solution (e.g., we cannot understand why a rise in temperature is bad if we do not understand the consequences). At other times, when given a problem to solve in the form of a solution (e.g., implement cost reduction), we decide that it is best to comply with the assignment and please our boss rather than expend the time and energy to uncover and resolve the underlying malfunctioning relationships that are the real cause of disappointment.

This misdirection is not necessarily due to an inability to problem solve, or even a lack of intelligence, but without a functional understanding, the problem is often too elusive to be acknowledged as existing. The facts, conditions, and dimensions are not defined well enough to help managers make good decisions, yet we push ahead making ill-considered choices. We tend to put off some problems in the hope that they will go away, and so work on other problems that are better defined. In the meantime, the original problem gets worse. The process can go around and around, resulting in frustration because symptoms pop up all over the place and yet the problem won't go away. As the problem grows, more people are affected. Their added complaints describe how the problem affects them, adding yet more symptoms of the underlying malfunction, which is spreading its negative effects. Those same symptoms can appear to be different, depending on the people, disciplines, or departments affected because functions and the solutions that they perform are assessed in different ways (e.g., the accounting department notices rising costs, the planning department becomes concerned about schedule slippages, and quality control personnel see increasing amounts of rework needed, but no one see these effects stemming from the same cause, so they set about treating three symptoms of the same malfunction. This all makes the problem fuzzier until we decide to reduce the ambiguity and uncertainty between what we think is happening or should happen and what is really going on. That is, we must investigate and understand the fundamental causes before considering solutions.

The best way to define a problem in such situations is to describe it in writing. The description shows how we see the problem at that particular moment: how it affects us; how it affects others; the unpleasant situation that the problem causes (or the opportunity it is expected to capture). It is as Alice observed in Wonderland: "How can I know what I think until I see what I say?" At the Creative Problem Solving Institute in Buffalo, New York, this enigma is termed the *fuzzy problem*. The complex interrelationships result in a problem that tends to be too fuzzy or elusive when it is mulled over in the mind. Once a decision is made to attack a problem by taking it out of our minds and making it explicit on paper, the mental juggling that may have been going on for days, weeks, or even years can be stopped.

Setting Up the Problem in the Fuzzy Problem Technique

When looking down a microscope, clarity is achieved by turning the lens to reduce the blurred views of what is being observed. A similar process occurs when setting up a problem for analysis. We begin by expressing the problem informally so that the description focuses on the problem and not on the spelling, sentence structure, punctuation, and other distractions. The description should consider the who, what, when, where, why, and how aspects of the problem.

As a minimum, the description should address the following questions:

- (If personally involved) How do I feel about the problem?
- Who or what is affected?
- What makes this problem better or worse?
- Why do I consider this a problem?
- What are the consequences if not solved?
- Where did the problem originate?
- Who owns the problem (who is most concerned that the problem exists)?

Once a problem is committed to writing, a review of its conditions will disclose the fact that it is composed of a number of subproblems often working systemically. To illustrate the problem-solving technique, consider the following hypothetical situation.

Problem   We have seen several mice in the house during the past few months.

Description   Mousetraps have killed a few mice, but other mice are still around. We were scared the other day when my wife saw one near the baby's crib. There is enough concern about the baby's health without worrying about these mice. If we can't get rid of them, someone may get bitten. We usually don't see them during the day, but we hear them at night. The worst time was after a party we held recently. We didn't take the trash out until the next morning. When we got up in the morning, we saw three of them in the kitchen. We decided that if we can't get rid of them, we will have to move.

From this description, *subproblems* can be identified:

- Mice represent a danger to health.
- We are afraid of mice.
- We are concerned about the effect of mice on the baby's health.

By breaking out the subproblems (i.e., articulating multiple foci of consideration), the divergent process has begun. We are now removing our attention from the problem as perceived originally and enlarging our field of view. The list of subproblems can be increased beyond those listed originally in the problem description. Once we become aware that the problem has many dimensions, it is easier to consider the subproblems systematically in a way that avoids being overwhelmed by complexity.

A restatement of a particular subproblem into a creative question (How might ...? or In what way might ...?) will help us to see the statements in terms of an opportunity for a solution. *Problem statements* could be developed from the subproblems of the example, such as:

- How might I control and get rid of the danger of mice?
- How can I overcome my fear of mice?
- How can I protect the baby's health?

The original problem has now been structured into several creative questions. From this list there may be one or many problems that stand out as being more important to resolve than the others, including the originally perceived problem. The perceived malfunctions selected as being the most significant provide a new direction for identifying the main problem.

Assuming that the first problem statement (... danger of mice) was identified as the most significant statement of the problem, we now want to determine whether the scope of the problem should be broader or narrower, a step known as *problem scoping*. In problem scoping, a narrower problem is determined by answering the question *how*, and a broader problem is determined by answering the question *why*. The following shows the evolution of the *how* and *why* questions to the problem statement.

- *Question 1*: How might I control and get rid of the danger of mice?
- *Method 1*: Get a better mousetrap (trap mice).
- *Question 2*: How might I trap mice?

Note that in Figure 2.1, problem scoping stopped and speculation began. The transition from problem scoping to speculation usually occurs when the problem becomes so narrow that specific solutions will satisfy the *how* question. The key question here

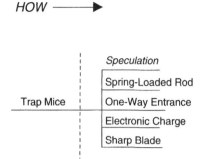

FIGURE 2.1   Trap Mice

is: Does the team really have sufficient understanding to move into solution-seeking mode? For the purpose of most value studies, adequate understanding is decided by time constraints as more and more clients want faster and faster results. This is being challenged by many in the VE world on the grounds that the goal is not to perform the act of thinking clearly as quickly as possible by degrading FAST models but to shave years off R&D cycles by spending a few days thinking more clearly than rivals do.

However, as demonstrated in the illustrations that follow, the option to speculate on specific solutions to a problem rests with the problem solver and his or her determination that the solutions will resolve the problem issues satisfactorily. Models reflect the quality of thinking that guides their development.

- *Question 3*: Why do I want to trap mice?
- *Intended outcome*: To eliminate mice
- *Question 4*: How can I eliminate mice?

Alternative methods to *Trap Mice* are *Obtain Predator* or *Remove Trash* or *Poison Mice* or *Sterilize Mice* or . . .

As shown in Figure 2.2, many more possible solutions flow from question 3, which yielded an intention of *Eliminate Mice* than flowed from question 1, which led to the specific outcome *Trap Mice*. The methods of trapping mice bring a smaller number of mechanical devices to mind, stimulated by the active verb *trap*. If *eliminate* is accepted as the active verb, the scope of the problem is enlarged. Since *Trap Mice* is only one option available to *Eliminate Mice*, by enlarging the problem definition we also increase the number of alternative solutions. As we don't want to collect mice but to get rid of them, we must be clear in our purpose when thinking about the action described by the verb selected.

In Figure 2.2 the *how* question for *Obtain Predator, Remove Trash, Poison Mice*, and *Sterilize Mice* was left unresolved. Responding to the *how* question would probably result in speculating on how to implement those solutions similar to the way that *Trap Mice* was elaborated in Figure 2.1. The problem can be expanded further in the *why* direction.

- *Question 4*: Why do I want to eliminate mice?
- *Intended outcome*: To prevent disease
- *Question 5*: How can I prevent disease? (See Figure 2.3.)

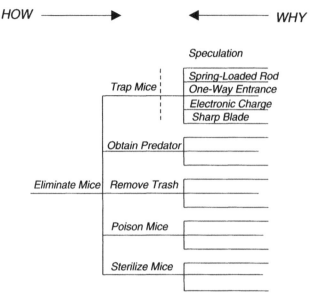

FIGURE 2.2   Eliminate Mice

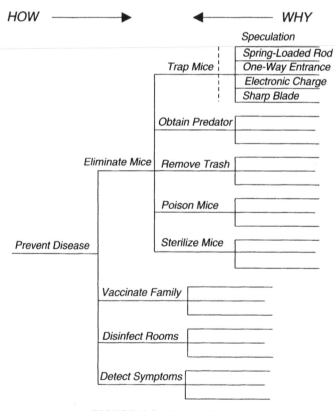

FIGURE 2.3   Prevent Disease

By continuing to ask *why*, we would expand the problem's linkage to other aspects of the tangled issues that need to be considered.

- *Question 6*: Why do I want to prevent disease?
- *Intended outcome*: To protect my family
- *Question 7*: How can I protect my family?

Generate alternative methods: *Prevent Disease* or *Install Alarm* or *Move Home*.

As the scope of the problem definition enlarges, so do the number of solutions available. By reviewing the problem scope, the real problem that needs resolution in order to stop other problems becomes apparent. We are making explicit our intentions by considering *how* and *why* in a logically structured way. It may be that the fundamental problem is preventing disease, and that the presence of mice, although annoying, is a symptom of the real problem. *Preventing Disease* would probably be the problem to solve in a subsequent phase since *Install Alarm* and *Move Home* to *Protect Family* (see Figure 2.4) is too abstract for the concerns expressed by the problem owner. If

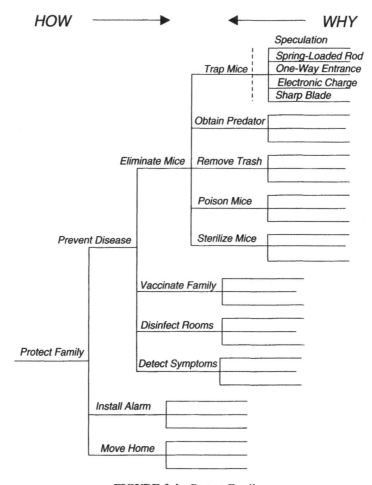

FIGURE 2.4   Protect Family

we were R&D directors and committed funds without working the process through, we may well have pursued the wrong problems.

The answers to the *how* and *why* questions are expressed as verb–noun functions to model the fact that we are thinking about intervening in a dynamic system. This naming convention will ensure that the problem remains at a high enough level of generality to encourage divergent thinking. The verb–noun suggests the action that possible methods or solutions must perform without prescribing them.

This technique is particularly helpful in identifying and scoping a problem that is not well defined, the type of problem that is abstract, where no clear path exists to find the real cause. The technique will help to identify key issues and focus on those issues to consider ideas and useful solutions. In the example above, the ways to *Prevent Disease* may have many solutions, not just one best solution. This is because the functions being considered are part of other functional relationships, such as *Feed Family* functions, drawn from the salary–economic system, and sometimes the best technological solution may be unaffordable and therefore inaccessible.

HIERARCHICAL TECHNIQUE

In the verb–noun function technique the function we were trying to accomplish by implementing a particular solution was identified. This goal of a system of functions is generally referred to as the *basic function.* The principal pioneer of FAST, Charles Bytheway,[2] argues that the basic function, sometimes called the *proper function*, is the only one that if removed would cause all other functions to become unnecessary. The word *basic* has more to do with fundamental importance than with simplicity. The functions *Produce Heat* and *Produce Flame* were basic functions that rely on a collection of other functions to be performed. The fundamental usefulness we are trying to identify with an artifact known as a cigarette lighter is its basic function. The purpose of the lighter is enabled by the performance of the basic function and is determined by users. In the example above of the fuzzy problem technique, *Prevent Disease* was the basic function selected, as *Install Alarm* and *Move Home* were not chosen to *Protect Family*.

VERB–NOUN AND FUZZY PROBLEM TECHNIQUES WITHIN THE HIERARCHICAL TECHNIQUE

Combining what we have learned so far in this chapter, we can see that other functions, performed by artifacts such as a cigarette lighter, must be carried out to make the artifact work. Like most products and services, the lighter is simply a system of functions that collectively enable the basic function to be performed. Among these other functions may be *Strike Flint, Discharge Electric Current, Rotate Wheel*, and *Activate Battery*. Obviously, one lighter does not perform all these functions. If we elected to *Produce Flame* with a spark from a flint, that lighter must perform the functions *Strike Flint* and *Rotate Wheel*. If we elect to perform the basic function with a battery type of lighter, that lighter would perform the functions *Discharge Current* and *Activate Battery*. The technology within the artifact is a measure of our knowledge and the options available to us to perform a basic function. The FAST model is thus a way of showing a team's knowledge in an efficient way.

These other functions that are often needed to make a method or solution work are termed *secondary functions*. Here the modeler must be sensitive to the distinction between a generic verb–noun function and a specific verb–noun process or solution. If we show *Rotate Wheel*, we have made a conscious decision to have a wheel that rotates in our model and thus limit our innovation scope. The practicalities of a particular situation must be seen as an influencing factor. Even so, the FAST modeler *must* be aware of the biases affecting the team's thinking. When placing verb–noun Post-It notes on an emerging FAST model, they should represent functions. If they are really nothing other than processes or other preconceived solutions, we might be inadvertently reducing the number of alternative solutions available for consideration. The modeling logic shows that secondary functions (whether generic or specific) are needed to cause a basic function to be performed. Sometimes we raise the level of abstraction from specific verb–noun representations (e.g., *Trap Mice*) to generic ones (e.g., *Protect Family*) so as to open the innovation opportunity. This is discussed again later in the book.

Once the basic function is agreed upon, it will not change regardless of the method selected to perform that function. This is because the basic function is about the use value of the object being innovated. What is more, customers usually purchase a product primarily to access the basic function, because of the worth of this usefulness. The concept of basic function is what gives an artifact such as an automobile or wristwatch its identity. The secondary functions generally change with the method selected to perform the basic function.

In representing a functional system from the basic function down through the detailed elements, there are many alternative solutions and subsolutions to consider and select. It is much the same as tracing a route through a family tree. There are many branches and limbs, and which ones we choose determine where we end up. What we are doing is uncovering the technological thinking that went into the artifact's original design. What is more, we are combining the causal properties of materials and objects in an intentioned system that we want to perform in order to achieve some useful outcome, such as work or aesthetic pleasure.

This method of problem identification is termed the *hierarchical technique* or *function family tree*. The hierarchical technique is used where there is a function relationship between the basic function and the method selected to achieve that function. The arrangement suggests a means–end relationship that unites the "big picture" with the "nuts and bolts" working drawings. To illustrate, we have a problem airlifting objects out of a congested area such as an inner city or a dense forest. The basic function is identified as *Airlift Objects*. The conditions of congested area, the size and weight of the objects, and other factors are constraints to be considered in later judgments. Figure 2.5 illustrates some ideas that were brainstormed to satisfy the basic function.

Each of these ideas has significantly different secondary functions to help make the respective concept work. For example, what functions would be necessary if *Hot Air Balloon* was selected? In evaluating which of these concepts best meets the constraints imposed on the problem (congested area, size and weight of the lift object), the idea of using a hot air balloon is selected. For the balloon to work, it must satisfy natural or casual functions to do with overcoming gravity. Figure 2.6 shows some of the *secondary functions* required for the hot air balloon. Each of these functions must be satisfied by solutions. There will be additional constraints imposed on these functions, how much gas to contain, what type of gas, how hot, how rapidly to heat, how responsive the

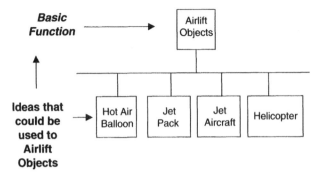

FIGURE 2.5 Hierarchical Technique and Airlift Objects

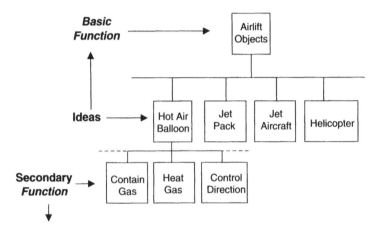

FIGURE 2.6 Secondary Functions

controls are. For each of these functions, ideas are generated. In Figure 2.7 the function *Heat Gas* will be pursued to show some solutions that might satisfy the function.

Note in Figure 2.7 that the term *New Basic Function* appears lower than the previous basic function *Air lift Objects*. Since we moved the focus of our thinking down a level of abstraction from the generic to the specific, thus making *Heat Gas* the key problem issue, this function becomes the basic function. That is, the decision to use *Heat Gas* as the key focus removes other functions from further consideration. If, as an example, we were in the business of making hot air balloons and wanted to develop or improve our method of heating gas, our problem would have started out at that level. We would not have started at the level of *Airlift Objects*. However, if we were responsible for airlifting the object in question, or owned the object, we would begin our problem analysis at the higher level, *Airlift Objects*. This selection of appropriate levels of abstraction and framing problems and opportunities is very important and is linked to value. This provides us with a clue, as identifying the customer's view of value often sets the level of the basic function.

As we work down through the hierarchy of solutions, the scope of consideration narrows. As the scope narrows, we elicit new basic and secondary functions for that level. Thus, we target our creative energies on functions that we wish to improve in

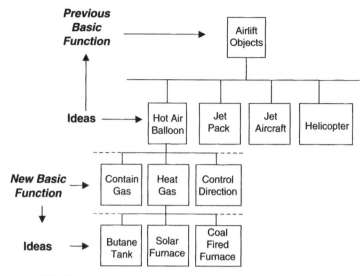

FIGURE 2.7 Functions, Ideas, and Levels of Abstraction

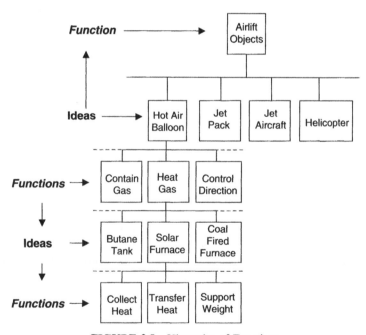

FIGURE 2.8 Hierarchy of Functions

the same way that your doctor considers what medication to prescribe to address some malfunctioning system in your body.

Let us assume that the concept of a solar furnace best met the constraints for *Heat Gas*; Figures 2.8 and 2.9 illustrate the new functions generated to satisfy that concept. Depending on the scope of the study, we can enter the model at any function level and work down. Wherever we enter, all solutions and secondary functions from that

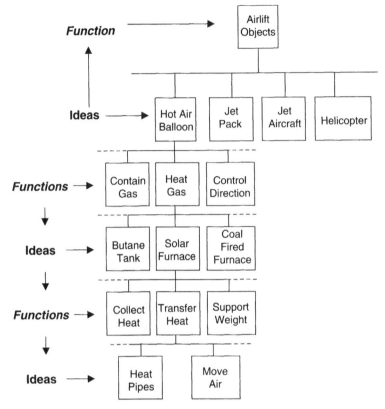

FIGURE 2.9   Hierarchy of Solutions

level down are subject to change. In many cases, a solution already exists but may not be working as well as needed and so requires a different approach. In such situations we recognize a function as being a malfunction because the solution to perform the function does not give satisfaction. A hierarchical diagram can be constructed that shows the functions and solutions as they exist.

A study of the diagram will be helpful in identifying the problem area(s) and then working from that point, or those points, down. Once the malfunction is identified, moving up through the higher levels could identify the cause of the problem and its symptoms (i.e., identify the cause of the dysfunction and dissatisfaction). The process can also supply clues to determine if seemingly unrelated symptoms stem from the same malfunction problem.

CLOSING REMARKS

We learned in this chapter that by focusing our descriptions on active verbs and measurable nouns we can build complex explanations of what work something needs to do. We also showed how a function is an abstract set or class of solutions rather than a specific or particular solution. This enables us conceptually to swap the way a function is performed and enables us to coordinate the innovative energies of a multidisciplinary team.

We also saw how the fuzzy problem technique enables us to move beyond superficial problem statements and get closer to basic causes. By spending time in such an enquiry, we can reduce our wasted energy and limited creative resources by targeting and dealing directly with fundamental causes. The hierarchical technique showed how we can move levels of abstraction and how that enables us to align our innovation process to value systems such as an organization's strategic ambitions.

The three topics cited above are not to be treated as disconnected or isolated from the remainder of the book. They combine and synthesize into the underlying theories that become the act of building FAST models. In the next chapter we examine function analysis in more depth, and that will set us up to discuss FAST modeling.

CHAPTER 3

FUNCTION ANALYSIS

Let us begin this chapter by exploring the relationships among value, function, and innovation using the notion of an elegant solution as our guiding proposition. When NASA realized that a ballpoint pen could not write in space, consultants were commissioned to invent one that could. Although we don't know the actual costs, let's say that it took $10,000 to get a group of engineers together to develop a prototype pen that could write in a zero-gravity space environment. On the next mission, someone asked how the Russians managed to write in space, to which the reply was: "They take pencils."

We hope you smiled at that story. We would now like to ask you why you smiled? Is it because you intuitively know that use of a pencil costing about 50 cents compared to a pen costing $10,000 constitutes a more elegant solution? But why is it an elegant solution? The expensive pen had an elaborate design with pumps and other components. So it had a far higher number of intrinsic functions than those of a pencil (an intrinsic function is one that resides within the artifact). Here, then, is a clue to an elegant design. An elegant design reduces the number of intrinsic functions to a minimum by efficiently selecting technologies that perform the basic functions within the weighted values that the evaluator sets (i.e., customers, not manufacturers, decide what counts as value). One clue to elegance is fewer essential components satisfying the same need. One has only to consider the need to sharpen a pencil in space to realize that the functionality required is larger than the pencil itself. Also, if the evaluator who weights the criteria selection has values that differ from those of the customers, the wrong selections could be made. Just as customers control revenue, managers control cost. So it is common to hear of cost-cutting drives when sales are flat, instead of trying to grow value.

Stimulating Innovation in Products and Services: With Function Analysis and Mapping,
by J. Jerry Kaufman and Roy Woodhead
Copyright © 2006 John Wiley & Sons, Inc.

Function analysis is the foundation upon which value methodology is structured.[1] It is the one discipline that separates VA, VE, and VM from the many problem-solving initiatives and processes available to the problem solver. The power of function analysis is to separate the intent or purpose of a thing from the description of its superficial appearance. It enables us to move from the concrete world of observable phenomena to an abstract world that facilitates creativity. However, whereas the technique enables FAST models to be developed, unless those models correspond to how the observable world works, we are not so much building representations as making modern art types of pictures. It is this ability to operate conceptual thinking with a systematic method that reflects reality and makes function analysis and FAST models so useful in innovation. Manipulating the way that functions are performed to leverage value is credited to Lawrence D. Miles, and we present some of his theories here before building on them to move closer to today's common practice.

FUNCTION ANALYSIS SYNTAX

Miles used the verb–noun discipline described in chapter 2, as the rule for expressing functions, and expanded the two-word description by using an active verb and measurable noun to best describe functions. We now build on Table 2.1 and consider this in more detail.

Active Verbs

When identifying the functions within components, products, or processes, it is important to use active rather than passive verbs (Table 3.1). We want to map out the work that must be performed within a system, so *action* is what it is really about. The verb describes the specific action we plan to carry out to achieve our intended purpose. The noun defines the object on which the action operates. Searching for the most descriptive verb–noun is difficult. Compromise often results in selecting the action as the noun and using a passive verb to complete the function description. We must never lose sight of the need to represent a function as a dynamic force that is essential to the working of that object or system in which it resides.

If a passive description of a function is suspected or you wish to express the function more actively, try to use the noun as a verb and then select another noun. For example, *provide* is commonly used when the contribution that a function makes to the system in which it resides is not fully understood. *Provide* is a word that lacks any sense of specific action necessary to achieve the basic function. Functions are intended to be taken literally as we attempt to describe <u>what</u> a system actually does in terms of <u>how</u>

TABLE 3.1 Passive Versus Active Verbs

Passive	Active
Provide support	Support Weight
Seek approval	Approve Procedures
Develop exhibits	Exhibit products
Submit budget	Budget expenses
Determine resolution	Resolve problem

and <u>why</u>. To use the verb *review*, as in the function *Review Proposals*, means read or skim but not to comment or take further action. A precise meaning is being referenced. If this is the intent of the function, the function is stated correctly. Use of the verb *attend* will have different meanings depending on the conversations in which it is used. For a staff member to *Attend Meetings* means that the person is just expected to sit there. But for a nurse to *Attend Patient* means that the nurse is expected to care for the patient.

We need to be clear about what we mean when selecting words and their meaning that flows from the rich dialogue which surrounds the construction of FAST models. Using *ize*, as in *prioritize, modernize, monetize*, and *economize* is shorthand for a description of activities. Although initially, the practice of using *ize*-ending words was discouraged, their use is becoming common practice as such words appear in dictionaries. Wherever possible, avoid using the verbs *provide, review, attend*, or verbs ending in *ize* so as to better capture the intended actions in the articulation of how a system works. The goal is to achieve a precision of thought that yields insights. We want clarity of message that allows groups of people to build team knowledge.

Measurable Nouns

Measurable nouns are easier to determine when the study topic is an artifact such as a cigarette lighter. When tangible *hardware* components are used, measurements are usually quantitative and often expressed as engineering units. Examples of measurable nouns include *weight, force, load, heat, light, radiation, current, flow*, and *energy*, to mention a few. They are from the scientific world to which an engineer is introduced at the university, a world full of causal actions and reactions. Functions such as *Control Flow, Reduce Weight*, and *Transmit Torque* have nouns that can be measured universally. Hardware system functions such as *Repair Damage, Complete Circuit*, and *Store Parts* have nouns that can be measured quantitatively but do not fit readily into a conventional lexicon of measurable nouns. *Damage* can be measured in terms of cost or time to repair; *circuit* can be measured by the size of the network or the energy consumed; *parts* can be measured by quantity or dimensions.

Selecting the appropriate measurement for artifacts depends on the problem to be resolved. In nonphysical or intangible applications such as business processes, such functions as *Transfer Responsibility, Create Proposal*, and *Develop Plan* can be measured in terms of time, people, or qualitative measurement units. Again, the problem to be resolved will determine what measurements to use. Using measurable nouns to describe functions is important in evaluating and selecting the best proposal alternatives to resolve the problem issues and present the proposals for approval and funding authorization. The reason we bring in such a discipline is to make us think very clearly, yet avoid limiting innovation. Try to be consistent in the use of a noun. To use many words to describe the same thing simply introduces unnecessary ambiguity when we want clarity.

Table 3.2 builds on Table 2.1 and shows a list of commonly used verbs and nouns, divided into *work* and *sell* categories. Note that the *work* listed nouns are divided into *measurable* and *nonmeasurable* nouns. *Measurable* refers to conventional units of measurement that are understood universally. Subjective nouns (e.g., firmness, delight) can be measured indirectly with ordinal scales and nonparametric statistics. How we specify the metrics for more qualitative constructs depends on the problem to be resolved and the goals to be set and met or surpassed.

TABLE 3.2   Common Verbs and Nouns

Verbs		Nouns	
		Work Functions	
		Measurable[a]	Nonmeasurable[b]
Support	Change	Weight	Component
Transmit	Interrupt	Torque	Device
Hold	Shield	Load	Part
Enclose	Modulate	Heat	Table
Collect	Attract	Flow	Damage
Insulate	Emit	Radiation	Circuit
Protect	Repel	Current	Repair
Prevent	Filter		
Amplify	Impede		
Rectify	Induce		
		Sales Functions	
Increase		Beauty	Form
Decrease		Appearance	Symmetry
Improve		Convenience	Effect
Create		Style	Loops
Establish		Prestige	Exchange
Enhance		Features	Cost

[a]Measurable in engineering units.
[b]Other units of measure.

Many functions aimed at increasing sales volume are considered secondary, but they are of primary importance if they contribute to the sale of the product. After all, in many commercial settings the objective is to improve value as determined by the buyer, not simply to reduce cost. As the customer controls the revenue stream, to assume that lowest cost is the only solution is to have a very narrow view of what sways a customer's purchase decision. In many high-technology industries the late arrival in a market loses far more than a 20 percent cost reduction could gain, as "first to market" sets the price and other customer expectations. We must always remember that a function is that which makes a valuable contribution to the system in which it resides. That contribution is assessed through the particular solution chosen to perform the function.

Using Two Words to Describe Functions

Using two-word function descriptions is essential because it cuts through technical jargon and creates a clear communication format that allows members of an interdisciplinary team to communicate with each other. It allows scientists to communicate with financial analysts, engineers with procurement, and manufacturing with marketing and so provides a very useful communication role in multidisciplinary and cross-functional teams.

To cite an example of garbled communication, if the finance representative on our hypothetical interdisciplinary team presented an idea for consideration, he or she might say: "Give consideration to obtaining our product at the present time while deferring actual expenditures of capital to a future period." In time, after some questioning, the suggestion would be understood. Using the verb–noun approach, this idea could be expressed as "Buy now, pay later." Although the many subtleties of finance might not be apparent immediately, the team has a better understanding of what is intended and can agree on what actions to consider.

Selecting two words, an active verb and a measurable noun, may sound like a simple procedure, but the opposite is true in practice. Lawrence D. Miles, the creator of value analysis, recognized the difficulty, and sometimes frustration, in trying to find those two elusive words that could most accurately describe the function of the item under study. In stressing the importance and difficulty in identifying functions succinctly, Miles[2] said: "While the naming of functions may appear simple, the exact opposite is true. In fact, naming them articulately is so difficult and requires such precision in thinking that real care must be taken to prevent abandonment of the task before it is accomplished."

The team must confirm their understanding by arriving at a consensus of the function description. If the verb–noun label suggests what <u>really</u> happens, we can agree. If it is not so obvious, our understanding lacks the precision born of clarity, and the conversations should continue.

DEFINING AND CLASSIFYING FUNCTIONS

The word *function* is commonly used and has many definitions. For our purpose a function is defined as *"a necessary intent or causal action that is realized through the performance of a solution."* The two operative words in this definition are *intent* and *realized.* How a product or service is used does not necessarily identify its functions. A heavy book may make an excellent doorstop, but the intended function of a book is not to *Prevent Movement.* The intended function is to tell a story or to present information about the essential contribution being made in a valued way. The use of a thing's properties for practical advantage is why FAST models aid in innovation.

In developing value analysis, Miles further classified and defined functions to configure a system where functions can be separated from their physical parts. Once defined, functions can be examined and analyzed to determine the item's contribution to the *value equation*:

$$\text{value} = \frac{\text{function}}{\text{cost}}$$

Customers buy products and services for what they achieve. This can be for mechanical benefits such as a television set, aesthetic benefits such as a work of art, or combinations of these two types of benefits. The benefits are the product of solutions that are made relevant by the functions performed. That is, customers don't simply buy a product that they purchase functionality. The difference in cost between rival products is a reflection of the ingenuity used by the producer of the product. This can be summarized as a relationship between the outcome (i.e., value) and two key variables: the function and the cost of the solution. The difference between a great product and a poor product can be approximated through the foregoing equation. A great solution will perform more functions that customers want, with solutions that reduce the number of parts in the product and for less cost. We must be careful to remember that value is not our judgment but that of the customers. The value equation is the way that Miles expressed the relationship between function and cost. Miles stated that "all cost is for function." In a broader sense, *resources* can be substituted for *cost* and the relationship to value would still be valid.

In the context of an isolated examination of something taken out of a working entity, it is easy to view value in a detached way. The weakness of such a perspective is that in an isolated examination of a single part, the contribution to overall value may be counterintuitive. For example, a single part with a function performed by an expensive solution may lower the overall cost of the system or machine in which it resides.[3] Linking the functions' relationships together minimizes this weakness. In Chapter 4 we explore FAST modeling as a technique that helps.

TYPES OF FUNCTIONS

Let's now recap, and as we do so, move a little deeper into our exploration of functions. The classifications of functions we have been discussing are defined here.

Extrinsic Functions

Extrinsic functions, although outside the system under study, are necessary for the purpose to be achieved. For example, if we were manufacturers of television sets, the extrinsic functions would be related to how users control the operations of a TV set's internal functioning, such as volume levels and channel changes. These are commonly used in an approach called *customer FAST*,[4] in which the outcomes to be achieved are elicited directly from customers.

Intrinsic Functions

Intrinsic functions are those within the scope lines and to the right of a means–end major logic path. These are the functions within the system that we are studying which must be carried out via solutions in order to perform extrinsic functions. When the TV user presses a button on the remote control, the intrinsic functions do the necessary work to achieve user satisfaction.

Basic Functions

A basic function is defined as the principal reason(s) for the needed existence of a product or service operating in its normally prescribed manner. This is usually the first intrinsic function named on a major logic path within the scope of the study, as shown in our discussion of the fuzzy problem technique in Chapter 2. It is the function that manufacturers and customers use to attach an identity to artifacts such as an oven or an airplane and so is representative of all types. If we ask what an airplane does, the response would capture the basic function in semantics, such as *Transport People* or *Transport Cargo*.

Naming functions with a verb and a noun is what forces us to push aside the super-ficial meanings that we give to words to find an objective reflection of what a thing does that makes it a useful example of a certain type. So important is this clarity that manufacturers, especially in Japan, use the basic function to test the demand for a product in the preliminary stages of market research and analysis with a variant of function analysis called *quality function deployment*. The word *basic* is not meant to suggest a lack of technological complexity; rather, it suggests a quintessential usefulness that we expect from a certain type of artifact and is why a light is a light and a car takes you from A to B.

Secondary Functions

The other intrinsic functions that make possible the basic function(s) are secondary, but their connection through means–end logic signifies that they, too, are necessary. This makes them sort of subfunctions that need to be performed if we are to realize the basic function. These secondary functions are the way that products are configured and used to differentiate standard from luxury products. As secondary functions can distinguish a low-cost artifact from a high-quality artifact, we must be aware that aesthetic functions do work as well as mechanical use functions.

Practical Definitions

As you become more proficient in FAST modeling you will want to categorize certain types of function so that the explanation of the team's view is clearer and less ambiguous. Many value engineers prefer a simplified definition, which describes a basic function as *anything that makes the product work or sell*, and those functions that do not support this definition are classified as secondary functions. Secondary functions are sometimes subclassified as *required functions*; these are almost extrinsic functions, as they reside outside the functioning of the system but are still to be achieved: for example, the color of a car as used to increase sales volume. Required functions describe an external function (e.g., of individual preferences) or some feature that may not contribute to the means–end logic of a basic function but is mandated by a customer (e.g., children's medication that tastes nice). The size and layout of a personal computer keyboard, the incorporation of an emergency brake in automobiles, and placing buttons on the sleeves of men's sport and suit jackets all represent costly functions that do not necessarily contribute directly to the basic function but are mandated by the customer. Eliminating or modifying solutions used to perform secondary functions would not necessarily affect the product and may even improve it. However, not satisfying these requirements would result in lost sales because customers, not management, decide value and control revenue streams.

RULES GOVERNING BASIC FUNCTIONS

Let us look at basic functions in more depth. It is important that we capture the essence of a basic function in a way that helps us to recognize how customers would see usefulness. With the aim of logical coherence, we present four rules that govern the identification and intended behavior of basic functions and are important when selecting and classifying a function as basic.

- *Rule 1*: Once defined, a basic function cannot change, as it becomes the principal work that the system of intrinsic functions is expected to do. For example, a flashlight is at a minimum expected to emit light or the concept of a handheld torch is meaningless.

The importance of this rule is that those functions designated as basic represent the work function(s) of the item or product and must be maintained and protected. Determining the basic function of single components can be relatively simple. The basic function of a spring is to *Store Energy*; a screwdriver *Transmits Torque*; a screw *Joins Components*. As components are joined together to become assemblies, and assemblies are combined to create products, determining the basic function of the combined components becomes more complex. Is the basic function of a cigarette lighter to *Create Heat* or *Produce Flame*? As discussed for verb–noun function techniques in Chapter 2, the answer depends on the focus of our study and how that frames the problems or issues to be resolved. If the company designing a new cigarette lighter has economic ties to the production of lighter fuel, *Produce Flame* would be the obvious choice for the basic function.

We must be clear that economic value is achieved by designing use values that customers are prepared, and are able, to exchange money for. Preferred functions are those the customer is willing to pay for. If the designers of the cigarette lighter are a new venture company without invested capital, inventory, and other economic constraints, the basic function could be *Create Heat* because *Create Heat* gives designers more conceptual freedom than *Produce Flame*. The same rationale applies when identifying the basic function of a pencil. Is it to *Make Marks* or *Deposit Graphite*? To *Make Marks* offers more creative potential than *Deposit Graphite*, which sounds more like a solution than a function to be performed and as such narrows the range of creative thought. You should determine the problem to be resolved before selecting the basic function, and this must be thought of in terms of economic priority as we try to influence customers to increase our revenue streams. By definition, then, functions designated as basic will not change, but the way those functions are performed can be changed.

- *Rule 2*: The cost contribution of a basic function is usually a minimal percentage of the total product cost.

Drawn from empirical observations over the years, the importance of the basic function to the success of any product, and the cost to perform that particular function among all the other intrinsic functions, depend on the number of secondary functions that are or are not needed. Here we refer back to the discussion involving the NASA space pen costing $10,000 and the Russian pencil costing 50 cents to underline the link between rule 2 and an elegant engineering solution. This relationship is especially true in consumer product markets, where there seems to be an inevitable desire to add unwanted functionality to products for the sake of differentiation. We must define value and worth from the customer's perspective. Value engineering defines worth as "the lowest cost to perform a function reliably." The basic function of a $1 disposable lighter, *Produce Flame*, can be achieved reliably with a match costing less than one-tenth of a cent. The basic function of a Rolex watch, which costs in excess of $20,000, can be expressed as *Indicate Time*. A no-name, blister-packed wristwatch displayed on a self-serve rack in a drugstore, costing less than $5, will perform the basic function *Indicate Time*. Why do people purchase Rolex watches? Certainly not for their ability to perform the function *Indicate Time*. People buy Rolex watches to show other people that they can afford to buy Rolex watches. Creating a perception of success by possessing a Rolex is the primary motivation for the purchase. However, if that $20,000 watch did not work and could not be repaired, in terms of function loss the value of *Indicate Time* is $5 out of $20,000. Could you then say that the watch is now worth only $19,995?

Of course not. The $20,000 price tag is based on the expectation that the watch will do what all good watches are supposed to do. This leads us to the third rule.

- *Rule 3*: You cannot sell supporting (secondary) functions without performing the basic function satisfactorily. For example, a luxury car that does not start is not of much use.

Most consumer products are not purchased because of the performance or the lowest cost of basic functions alone, as in the Rolex watch versus the disposable watch discussed above. The basic function, such as *Indicate Time*, is an unquestioned expectation. Whatever quality and price a car is, it has to do that core function which makes it a "type," in this case a type of car and not a type of washing machine. When purchasing a product it is assumed that the basic function works. The basic function is a fundamental expectation: The product is supposed to work. The customer's attention is then directed to those visible secondary support functions or product features that influence the perceived worth of the product.

From a product design point of view, products that are perceived to have a high value must first address the basic function's performance and core reliability, then stress the achievements of all of the other performance attributes that flow from secondary functions. The secondary functions are necessary to attract customers who want to distinguish between types of products, such as shoes, TV sets, and so on. Secondary functions are incorporated in the product to support and enhance the basic function and help sell the product. The elimination of secondary functions that customers do not value, or are not customer sensitive, will reduce product cost and increase value without detracting from the worth of the product. Here we use *value* as a means of preference, as in choice theory, and we use *worth* as a measure of economic exchange. Modifying functions that customers prefer will change the customer's value perception of the product, which could have positive or negative effects on sales.

- *Rule 4*: Loss of the basic function(s) causes loss of the market value and worth of a product or service. For example, an irreparable Rolex wristwatch that can never display the correct time cannot be sold for the same price as one that works. Its sale price would be considerably lower than that of a comparable model, even though the motivating factor to purchase the Rolex was *not* the common basic function that groups all wristwatches.

In differentiated products, the cost contribution of the basic function does not, by itself, establish the value of the product. It is quite simply a basic or fundamental expectation that we require a certain type of thing to do as implied in its identity. Few products other than commodities are sold on the basis of their basic function alone. Although the cost contribution of the basic function is relatively small in most products, its loss, as illustrated in the lighter and watch examples, will cause the loss of the market value of the product.

In the lighter example, when it no longer performs its basic function, the lighter is discarded. Why? The lighter has about 17 mechanical parts, all of which continue to function. The valve is still operable, the flint can still produce a spark, the springs continue to store energy, and so on. The reason the lighter is discarded is that it has lost its basic function ability, thereby losing its market value. The components of the lighter performed secondary functions that supported the basic function of the lighter.

Although each working component still functions on its own level of abstraction, the parts lost their value as supporting functions when the basic function failed.

FUNCTION IDENTIFICATION EXAMPLE

In Figure 3.1, a number of function descriptions are given for a common electrical fuse. All of the functions describe what must be done for the fuse to be useful. The reader's assignment is to select which of those functions offered best describes the basic function. Do any of the functions describe the basic function accurately? If not, can you think of one?

If you selected *Break Circuit* as the basic function, consider what would happen if you removed the fuse and did not replace it in the circuit. The fuse has some usefulness to do other work in addition to *Break Circuit*. How about *Connect Circuit*? If this is the basic function, why not eliminate the fuse from the circuit and connect the wire ends? Again, the fuse has some usefulness that makes the choice *Connect Circuit* inappropriate. Let's consider *Protect Equipment*. If this is the accepted basic function, ideas such as a security fence, guards, and guard dogs could be considered as alternative solutions in brainstorming the basic function *Protect Equipment*.

A better description of the usefulness that a fuse provides would be to protect electrical equipment by preventing electrical surges from rising above the circuit tolerance, entering the equipment, and destroying unprotected components: It's about the essential contribution necessary for a system to work. Remembering the lessons regarding fuzzy problems in Chapter 2, we could ask: How is that accomplished? The answer would provide the function: to *Limit Current*. We can now see that the usefulness a fuse provides is the basic function *Limit Current*.

For your next assignment, determine some functions for each of the three common items described in Figure 3.2 and select their basic functions.

Hints:

- The basic function of an oil filter is *not Clean Oil*.
- The basic function of a screwdriver is *not Drive Screws*.
- The basic function of a wall thermostat is *not Control Temperatures*.

Check the end notes for solutions.[5]

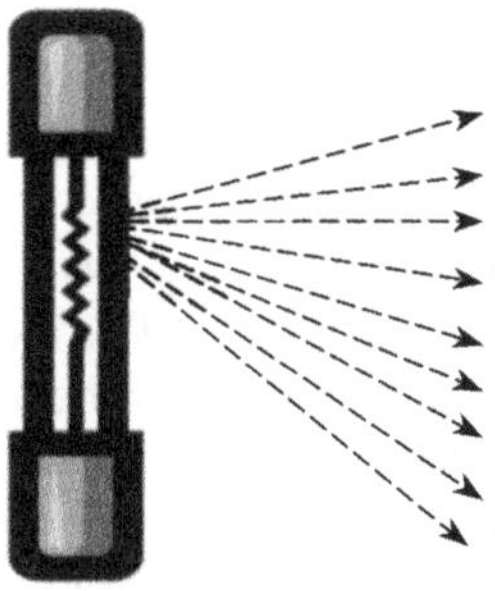

Random Functions	
VERB	NOUN
Break	Circuit
Connect	Circuit
Protect	User
Protect	Supplier
Protect	Equipment
Identify	Failures
Advertize	Manufacturer

FIGURE 3.1   Basic Function of a Fuse

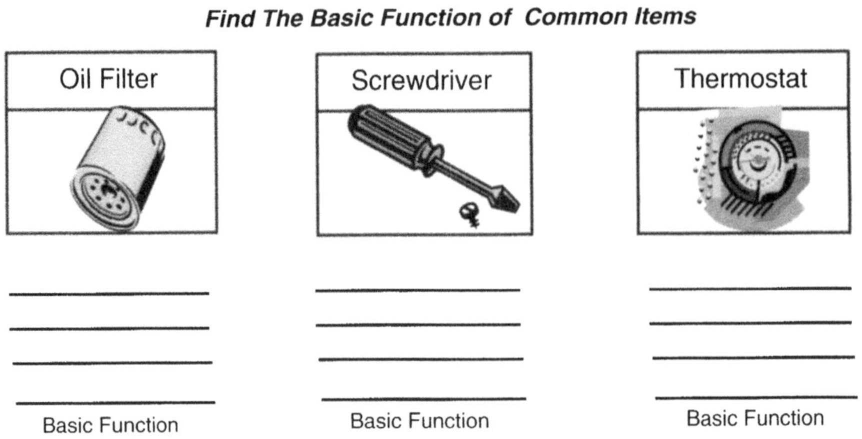

Find The Basic Function of Common Items

Oil Filter	Screwdriver	Thermostat

_____   _____   _____
_____   _____   _____
_____   _____   _____

Basic Function   Basic Function   Basic Function

FIGURE 3.2   Functions of Common Items

RANDOM FUNCTION DETERMINATION

The exercises just completed describe part of the function analysis technique created by L. D. Miles, in which we categorize function types. Now we explore how this is done in practice by looking at a technique called _random function determination_. In this technique we randomly select the items of a product, identify their functions, and then determine which of those functions are basic and which are secondary. The word _random_ refers to the starting point, not to the process itself. The reason we do this is so that we can better see which aspect of a product or service can be improved so that customers prefer our offering to that of a rival. We examine random function determination in more detail in Chapter 7.

Levels of Abstraction

As shown in the fuse example, a simple product can have many functions, among which can be found its basic and secondary functions. If we wish to examine the product in more detail, we can move to a lower level of abstraction and examine the components of the product and draw lessons from the hierarchical techniques discussed in Chapter 2. _Levels of abstraction_ thus refer to the level in which our thinking concentrates attention, such as at the general level, to a particular level, which could range from the entire universe to how oxygen and hydrogen atoms combine to form water. The fuse consists of a ceramic tube, a low-electrical-tolerance resistance strip, terminal ends, and a bonding medium to hold the product together. Each component of the fuse performs many functions. Examining each component on its own level of abstraction rather than as the machine _in toto_ yields the realization that each of the components often has at least one basic function that justifies its reason for being there, and if that is not the case, we could remove it without any ensuing effects. That is, each component usually has a single usefulness that gives rise to its reason for being there. However, when we look at all the components, many of those individual basic functions become secondary as we move the focus of the study to a higher level of abstraction.[6] The ceramic tube and terminal ends are important to the fuse but are less important when evaluating the contributions of the fuse to the equipment it supports. As we continue to move to higher levels of abstraction, the equipment (e.g., office computers) becomes less important to the system it supports (e.g., the computers in the sales department) and that system (e.g., the sales department) then becomes less important to the overall product (e.g., the enterprise itself), for which the fuse may be viewed as insignificant. Here we must be sensitive to our focus (i.e., scope of investigation) and the confusion we can create if we do not maintain consistency between the big picture and the nuts and bolts we are trying to represent as a system of functions.

Function and Component Selection

When a function analysis was conducted on a common lead pencil, the pencil's basic and secondary functions were determined by evaluating the components' contributions to the overall product (see Figure 3.3). The pencil has one basic function, *Make Marks*, and that function is accomplished by the lead in the pencil. If we do not need to *Make Marks*, all the other functions are unnecessary. Since all of the other functions are secondary, or in support of the basic function, they are candidates for elimination, consolidation, or modification to increase the efficiency of the product.[7] The manipulation of secondary functions is accomplished by the creative efforts of the team, provided that the team does not change the basic function of the pencil. This does not mean that a sintered rod of carbon graphite (or lead) must be used to perform the basic function. It does mean that at the conclusion of the pencil study, the pencil must retain its ability to *Make Marks* in a manner that satisfies the customer.

In random function determination, the manipulation of secondary functions is guided by reducing the cost contribution of performing those functions. The objective of VE, to improve value (i.e., preferences and worth) by reducing the cost-to-function relationship of a product, is achieved by seeking solutions that perform the basic function satisfactorily while eliminating or combining as many secondary functions as possible. In the pencil example, if you make a longitudinal cut in the wooden body and remove the lead carefully, you can write (make marks) with the lead without the benefit

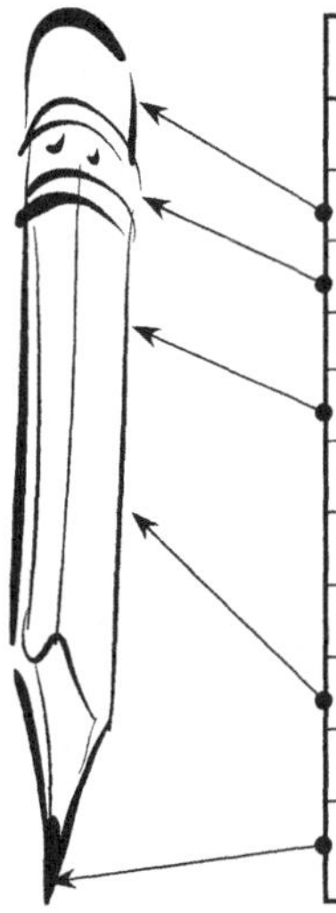

DESCRIPTION	FUNCTION	B	S
PENCIL	MAKE MARKS	X	
ERASER	REMOVE MARKS		X
BAND	SECURE ERASER		X
	IMPROVE APPEARANCE		X
BODY	SUPPORT LEAD		X
	TRANSMIT FORCE		X
	ACCOMMODATE GRIP		X
	DISPLAY INFORMATION		X
PAINT	PROTECT WOOD		X
	IMPROVE APPEARANCE		X
LEAD	MAKE MARKS	X	

FIGURE 3.3 Functions of a Pencil

of any of the secondary functions. Does this represent the best value proposal for the pencil? Best value is achieved only if the proposal is successful in the marketplace. It is in the marketplace that customers who can choose among many offerings select either your product or a rival's.

Function Cost Matrix

The function cost matrix approach to performing function analysis is a graphical extension of the random function determination method that we touched on briefly a few pages ago. The objective of this process is to draw the attention of the analysts away from the cost of components and focus their attention on the cost of the current solutions selected to perform the functions (see Figure 3.4). The function cost matrix approach displays the components of the product and the cost of those components along the left vertical side of the matrix. The top horizontal legend contains the functions performed by those components as determined in a random function determination exercise. Each component is then examined to determine how many functions that part performs and the cost spent on those functions. As an example, the eraser cost 43 percent of a penny, and all of that cost is dedicated to the function *Remove Marks*. At the bottom we see that the current solution to *Remove Marks* cost 0.43 penny, and that is 15 percent of the total cost. The metal band, which costs one-fourth of a penny, performs three functions: *Secure Eraser, Improve Appearance* of the pencil, and *Transmit Force* when using the eraser. In a cost estimate, we roughly allocate the cost to perform these functions by estimating the process and material cost of the metal band component.

To determine the cost of the aesthetic function *Improve Appearance*, the illustration shows that the metal band, body, and paint contribute collectively to that function. The analyst can now read the chart vertically to determine the cost contribution of performing the function *Improve Appearance*. At this point in the process it is more important to determine the approximate cost of the functions than to attempt to determine accurately the actual cost allocated to each function. Detailed cost estimates become more important in function analysis when evaluating alternative solutions.

Reading across the row marked "total" will show the cost and percent contribution of the functions of the pencil. This will guide the team during analysis to identify malfunctions that require further improvement. Here we are directing our innovation and creative resources to those areas that will yield most value. In this example 56 percent of the pencil's cost is dedicated to performing two functions, *Make Marks* and *Transmit Force*. If we were more concerned with improving appearance, a different innovation agenda would be drawn up.

Simplifying the Process

At this point you may surmise that the complexity of a product or system that contains a great many more components than a pencil would be substantially higher. Put simply, after going through the pencil exercise one could not begin to imagine what a function cost matrix of an automobile would look like. However, as we alluded to when we discussed our focus, complexity is not so much governed by the number of components in a product but by the level of abstraction that we, the observers, select to perform the analysis. In our automobile example, a high level of abstraction could contain the major subsystems as the components under study, such as the power train, chassis, electrical system, and passenger compartment. The result of a function cost matrix analysis could

Pencil Function Cost Matrix

COMPONENTS	COST (IN CENTS)	REMOVE MARKS		SECURE ERASER		IMPROVE APPEARANCE		MAKE MARKS		TRANSMIT FORCE		ACCOMMODATE GRIP		DISPLAY INFORMATION		SUPPORT LEAD		PROTECT WOOD	
		PERCENT	COST	PERCENT	COST	PERCENT	COST	PERCENT	COST	PERCENT	COST	PERCENT	COST	PERCENT	COST	PERCENT	COST	PERCENT	COST
ERASER	0.43	100	0.43																
METAL BAND	0.25			50	0.13	25	0.06			25	0.06								
LEAD	1.20							70	0.84	30	0.36								
BODY	0.94					10	0.09			40	0.37	5	0.05	5	0.05	40	0.37		
PAINT	0.10					50	0.05											50	0.05
TOTAL	2.92	15%	0.43	4%	0.13	7%	0.20	29%	0.84	27%	0.79	2%	0.05	2%	0.05	12%	0.37	2%	0.05

56% of Cost (MAKE MARKS + TRANSMIT FORCE)

FIGURE 3.4 Pencil Function Cost Matrix

focus the team's attention on the power train for further analysis. Moving to a lower level of abstraction, the power train could then be divided into its components (engine, transmission, driveshaft, etc.) for a more detailed analysis. Here a database could become a useful ally in managing the complexity. Another approach to simplifying the process while maintaining its validity is to select only the basic functions of the individual components for inclusion in the function cost matrix. This is a good approach for a simple product that has about 20 to 30 components.

As the focus of our study moves to a higher level of abstraction, many of the component's basic functions displayed in the matrix will be secondary to the product itself. For example, the basic function of a door, to *Control Access*, would be secondary to that of a house, whose basic function is to *Create Habitat*. To summarize, each component has a basic function. However, that basic function may be considered secondary to the system the component supports unless the basic function of the component is also the basic function of the system, as in our fuse example, where the basic function of the low-electrical-tolerance resistance strip is to *Limit Current*, which is the same basic function as that of the fuse.

CLOSING REMARKS

We revealed how the lessons in Chapter 2 are brought together in an explanation of function analysis. After explaining the concept of an elegant solution, we elaborated the verb–noun label used to identify a function and why active verbs and measurable nouns help innovation. We discussed some different types of function and the perspective of the modeler. If the modeler looks from a user's or customer's perspective, extrinsic functions are named. If the modeler looks from an engineer's or manufacturer's perspective, the internal working that enables extrinsic functions becomes the aim as we seek to understand intrinsic functions. Following this we explored basic and secondary functions and a few rules that make explicit some key relationships and expose some of the underlying theories upon which practice is founded. Then we looked briefly at how we move from a collection of components through random function determination to identify functions and whether they are basic or secondary, in order to appreciate how innovative thinking plays with these concepts and how we can seek either radical inventions or less radical ones by either challenging the basic function or restricting ourselves to considering only secondary functions.

CHAPTER 4

FUNCTION ANALYSIS SYSTEM TECHNIQUE

The techniques discussed in previous chapters build the foundation for FAST modeling. Understanding the principles that underpin those techniques will promote a greater appreciation of the power and universal applicability of FAST models.

PROCESS OVERVIEW

If you accept the premise that understanding a problem is the main part of getting to a solution, separating the problem from its apparent symptoms by analyzing its functions is essential to innovation. In that spirit we must see FAST modeling as our attempt to understand how a system should and does work in order to identify malfunctions, make a diagnosis, and recommend treatments. FAST models open the subject under study for analytical examination and the construction of a functional explanation of how something works. What makes FAST models powerful is that the conceptual breaking down into individual parts is contained within a format that allows a "joined-up" synthesis of the information embedded in the representation.[1] This forms a pivotal point around which a wide variety of subsequent approaches and techniques are included in the value methodology (e.g., VA, VE, VM). FAST models contribute significantly to perhaps the most important phase of VE: articulating theories of how things work.

The more contemporary use of FAST models is to complete the model because it is an excellent communication tool rather than stopping when the first idea has been achieved.[2] FAST models communicate clearly how a complex system such as an organization should work, in such a way that different disciplines get a collective view of intentionality and how causal relationships relate to those theories. A FAST model

Stimulating Innovation in Products and Services: With Function Analysis and Mapping,
by J. Jerry Kaufman and Roy Woodhead
Copyright © 2006 John Wiley & Sons, Inc.

should be seen as a dynamic model of a fully functioning system and used to generate insights, make diagnoses, and plan interventions through collaborative innovation.[3] Whatever the purpose for wanting to create a FAST model, its rules of logic must be applied consistently. Using the verb–noun rules in function analysis creates a common language that unites all disciplines. The verb–noun rule for identifying functions allows multidisciplined team members to cut through technical jargon, to contribute equally, and to communicate with one another while addressing problems objectively, without bias or preconceived solutions. FAST has also proven to be an excellent planning tool that makes the intentional *how–why* questions explicit and challengeable in a search for best solutions. Using the FAST modeling techniques described in this book will allow the accountant to communicate with the mechanical engineer and the physicist to explain his ideas functionally to the purchasing agent.

The two word verb–noun description of functions remains intact and focuses our analysis on "what is being done to what" to achieve another action, such as a customer buying our product. Distinguishing between basic and secondary functions and their associated function subsets is also incorporated into the FAST process. The most dramatic differences are in the use of more rules of logic to determine and test function dependencies and display the system graphically in a functionally dependent model.

The major difference between random function determination and the FAST process is in analyzing a system as a complete working unit rather than analyzing the individual components of a system in isolation from each other. When studying systems it becomes apparent that functions do not operate in a random fashion. A system exists because functions form dependency links with other functions, just as individual parts form a dependent relationship with other parts to make a system operate. The human heart functions so that circulating blood enables other kinds of functioning. It is this functional interdependency that links a component to other components as they collectively perform a system of functions when operating in, say, a machine or a living human body. FAST modeling is the product of team knowledge. It displays functional dependencies graphically and creates a process to study functional relationships that can be tested because they are explicit.

Since its introduction, FAST modeling has been used worldwide in a variety of areas, including marketing, engineering, manufacturing, new product development, revenue generation, cost reduction, hardware, and software. FAST has also been used in the medical profession and in state government agencies to create procedures for organization development and analysis and for productivity improvement and training.

In this chapter the subject of FAST models is explored more from a system and process approach than in its more common application to product development. The tendency for beginners using FAST models when studying hardware examples is to force-fit the FAST model to the product instead of questioning the functional relationships between the component parts and strategic outcomes. To the beginner, if the FAST model and the hardware project do not agree, it is assumed that the FAST model is wrong, the rationale being that the product must be correct because it exists and therefore the FAST model must be incorrect. The fact that the design configuration may not reflect the best approach to achieve its required functions is not considered an option to the beginner, causing him or her to lose confidence and shy away from using this most effective discipline. As an allegory, it is like the student who exclaims, "I tried using algebra but it doesn't work!" To avoid logic mistakes, we must rely on rules to guide us in our search for an elegant solution accessed by innovation and fueled by a comprehensive understanding of the necessary functionality shown in a well-built FAST model.

Paradoxically, hardware components and simple mechanical system examples such as cigarette lighters are the best way to teach the principles of FAST modeling. The beginner must relate something tangible to "visualize" the FAST modeling process. Remember that a FAST model is an abstract representation of the functioning performed by solutions as understood by the team building it. Having the right knowledge in the team is vital. FAST models give us an efficient way to map many suggestions and act as a reference point from which to coordinate the swapping of various solutions in an attempt to achieve improvement. A good introductory conversation is to identify a component or process and ask, "What's the point of having that there?" This leads the student to inquire about its functionality, which is conceptually distinct from the thing itself. As an effective management discipline, FAST modeling can be used in any situation that can be described functionally. However, FAST modeling is not a panacea. It is a way for teams to think together, but it also has limitations that must be understood if it is to be used properly and effectively.

FAST modeling was conceived initially as a system lacking dimensions. That is, it displays functions in a logical sequence, prioritizes them, and tests their dependency. It will not tell you how well a function should be performed (specification), when (not time oriented), by whom, or for how much; all a FAST model does is build a clear explanation of a collection of necessary actions upon certain objects arranged in a means–end relationship as understood by its authors, the team. But these chains of functions are normative. All we can ask is: How well does my current solution perform the functions? When we measure, we measure a particular solution that performs the function. This separation of function and solution means that we are free to consider alternative solutions, which are also measured to evaluate which selection offers the best value.[4]

The model can be dimensioned when completed, enabling the FAST model to be a useful means of coordinating and assessing the impact of planned changes. For example, if a FAST modeled an electrical circuit and the team wanted to consider different ways to *Limit Current*, alternatives to the fuse (discussed in Chapter 3) would be considered in relation to the value created by the entire circuit, and its importance. The old adage about not having a horseshoe nail, which leads to a domino effect that costs the kingdom a victory in war, is making the same point. We have to see value in context. Another adage talks of saving a penny's worth of tar and losing a ship. These stories describe decisions whose consequences can be avoided only by considering the impact at the aggregated system level. The lesson here is about seeking real improvement rather than short-term fixes that fail and cause more damage than existed before they were applied. Innovation is a product of changing solutions to perform functions within a system that combine to create greater value. A number of ways to modify and dimension the FAST model while protecting the integrity of the FAST process are explored and discussed in Chapter 5.

Some Misconceptions

FAST models will not solve problems in the sense that the process makes the solution conspicuously apparent. But it will help identify the cause of a problem, separating it from its symptoms and effects. FAST models are simply a representation of many ideas as held in the heads of the team building the model; the ideas are then laid out logically and tested collectively. The adequacy of the FAST model is a product of the team's knowledge and the clarity communicated by a neatly laid out model. The dimensioned FAST model will allow cross-referencing of functions to particular solutions in a logically arranged set of relationships that stimulate speculation on how else the functions can be performed for alternative solutions that improve value.

There is as yet no means to determine a correct FAST model, as in comparing it to some accepted norm, because it is a conceptual representation of the team's knowledge. There is a move toward valid FAST models using rules of logic. The degree of correctness and validity is directly dependent on the talents of the participating team members and how close the FAST model is to a true representation of the system being modeled. After agreeing on the problem statement, the single most important output of the multidisciplined team engaged in developing a FAST model is consensus. When one member says, "The function of X is Y," a discussion ensues that uses the collective intelligence and knowledge of the team, comprising diverse disciplines, to see whether or not the team believes this to be true. The role of external evidence and experiment can be used to strengthen the quality of the FAST model. This consensus must not be an imposed product of managerial power within the group but of how true the representation is. This is the difference between opinionated dictation based on power relationships and inquiring dialogue grounded in a respect for knowledgeable people striving for truth. Since the team has been charged with the responsibility of resolving the problem assigned, it is the usefulness of the FAST model in resolving the problem statement that is important.

Outsiders may comment and suggest revisions to the model (which may be valid contributions), but it is the team's model that must prevail because it represents a "buy in" by the team and is a reflection of their knowledge and understanding. When someone represents functionality incorrectly, the FAST model becomes a tool to stimulate organizational learning as others explain different perspectives. The team members must dialogue and reconfigure the FAST model until consensus is reached. All participating team members must be satisfied that their concerns are represented in the FAST model. They must also agree that the functions are expressed reliably and the solutions are practical. Consensus is seen as achieving an adequate explanation of how a system works. Then the team can move on to the next creative phase.

"As Is" Versus "Should Be" Models

There are two basic approaches to constructing a FAST model. The model can be displayed to show the current configuration of the product or process, called an "*as is*" model, or the FAST model can describe a conceptual, "yet to be developed" configuration called "*should be*." Which approach to use depends on the objectives of the study and the study's expected outcome and constraints. As an example, if a client desires an improvement in the existing solutions but wants to maintain the identity of the existing product, the "as is" approach should be selected. However, if the client desires a newly developed product or model, with few constraints such as backward compatibility considered, the "should be" approach is best. The major graphical difference between the approaches is in the complexity of the FAST model. An "as is" model displayed to, let's say, reduce product weight or cost must include a function description of the product, its subassemblies, and significant components in major and minor logic paths and activity terms (we explain these terms later in the chapter). This detail is necessary to establish a function–weight or function-cost relationship as the basis for the product improvement study.

The "should be" model is graphically much simpler because only the major logic functions are considered. By eliminating any preconceived solutions and thus many minor logic paths, a broader range for creative exploration is offered to the team in developing a new product or process. At this stage all you need to note is that there are some functions that belong to a major logic path and some functions that belong to a minor logic path.

SYNTAX USED TO CREATE AND READ A FAST MODEL

In this section we describe syntax to achieve consistency in creating and reading FAST models. It is important that the team creating the fast model understand how to read the model and communicate the model's knowledge. However, it is equally important that the model be understood when read without the aid of the model's creator. To achieve this objective of information networking requires language rules.

Reading *How–Why* and Our Intentions

In Chapter 2 we saw how a logic process using the *how* and *why* questions was applied to structure fuzzy problems. The directional references of the *how* and *why* questions discussed previously remain the same. *How* is read from left to right, and *why* is read from right to left. As an example, if we were addressing the function *Make Marks* and asked the question "How do we *Make Marks*?", the answer, in the form of a function, could be *Contrast Color* (see Figure 4.1). If we continued in the *how* direction (i.e., from left to right) and asked "How do we contrast color?", the answer could be *Deposit Medium* (see Figure 4.2).

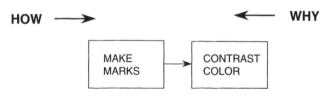

FIGURE 4.1 Building a Pencil FAST Model

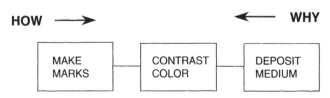

FIGURE 4.2 Deposit Medium

To test the validity of our attempt, we question the logic of the example and read the functions in the reverse *why* direction. "Why do we want to *Deposit Medium*?" To *Contrast Color*. Why do we want to *Contrast Color*? To *Make Marks*. If the team agrees with the theories, we can continue to expand the FAST model in either the *why* or *how* direction (see Figure 4.3). In the *why* direction we would ask "Why do we want to *Make Marks*?" "To *Record Data*"; and "Why do we want to *Record Data*?" "To *Retrieve Information*." Switching to the *how* question, we can continue to build in that direction by asking "How do we *Deposit Medium*?" "By *Apply Pressure*." Examining the function inputs thus far, the FAST model would look as shown in Figure 4.3.

If asked the question "What is the product represented by this FAST model example?" you answered "a pencil," this might not be correct. Consider the same function model and a typewriter. The model could be for a pencil or a typewriter or another means of performing the basic function. This shouldn't be too surprising because both products perform the functions described and have the same dependency order. The differences occur when the model is dimensioned in terms of time, legibility, productivity, or other measurements that reflect the problem under study. That is, when we expect a certain function to be performed in a certain way, we often influence the choice of the appropriate technology.[5]

How–Why Versus Why–How Orientation

The *how–why* orientation seems backward to many beginners of FAST modeling, particularly those involved in creating systems diagrams or process flowcharts. This is a valid observation if the principal objective is to create a flowchart, because the *why* direction describes FAST models in a systems orientation. There are a number of reasons for the choice of the *how–why* orientation.

1. Remember that means–ends logic shapes the dependencies, and when undertaking any task it is best to start with the outcome of the task and then explore methods to achieve that outcome. When addressing any function on the FAST model with the question *why*, the function to its left expresses the achievement of that function, or goal, which is itself a means to some other end. The question *how* is answered by the function on the right of the targeted function and is a reference to a method selected to perform that function being addressed, expressed in functional terms (see Figure 4.4). A systems diagram starts at the beginning of the system and ends with its output or outcome. A FAST model, reading from left to right, starts with the aim and ends at the beginning of the system that will achieve that outcome or goal.

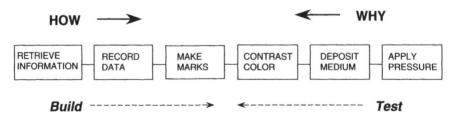

FIGURE 4.3 Extending a Pencil FAST Model

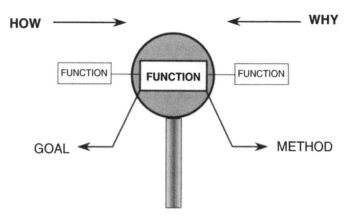

FIGURE 4.4   *How–Why* Orientation

2. Changing a function on the *how–why* path affects all of the functions to the right of that function, because a function to the right of another function is "dependent" on the function to its left, as they are in a means–ends relationship (see Figure 4.5). This is a domino effect that goes only one way, from left to right. Starting any place on the FAST model, if a function is changed, the ends are still valid (functions to the left), but the means to accomplish that function, and all other functions on the right, are changed.

Referring to Figure 4.5, if we didn't have to *Make Marks*, we might still be interested in *Record Data* (left), but would we be concerned with *Contrast Color* (right)? The marks could be replaced by an electrical signal stored magnetically and retrieved later. Functions to the right of another function are called *dependent functions* because the way in which a function links logically depends on the function to its left.

3. To find errors in the text, copy checkers often read a model in the reverse of conventional practice. In our case, reading from the "ends" or left side of the model, through the "means" to the beginning on the right-hand side (in the *how* direction), goes against intuition and the conditioning effect of the system paradigm. Because it

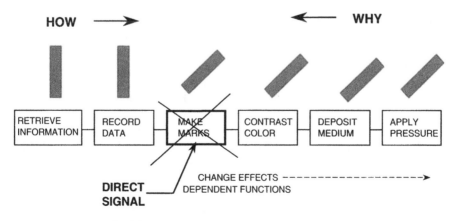

FIGURE 4.5   Dependent Function's Domino Effect

seems strange to build a model in the *how* direction (i.e., function justification), the team's attention will focus on each function element of the model. We deliberately make the team think in a different order to that which they are used to. Reversing the FAST model and building it in its system orientation will cause the team to leap over individual functions and focus on the system, leaving functional gaps in the explanation of how something works because they are comfortable in that mode of working and thinking. As with all forms of logic, the rules are intended to force us to think both rationally and clearly.

A good rule to remember when constructing a FAST model is to build in the *how* direction and test or validate the logic in the *why* direction.

Reading *When* to Consider Causality and Consequential Functioning

The *when* direction is needed to link enabling functions and consequential functions to the means–end relationships within the major logic path. The *when* direction describes activities, independent functions, and outcomes (e.g., *Satisfy Customer*) that are the result of selecting a particular approach to implementing the basic function and other major logic path functions. In some cases a function must be completed to enable the desired function to be performed; think about the need to plug a TV set into the domestic electricity supply to enable the set to fulfill its function. Alternatively, there are cases when the technology used to perform a function leads to other consequential functions becoming necessary. Consider the need for a fan to cool the inside of a personal computer as the functioning silicon gets hot in performing its computational operations. As can be discerned from the above, the *when* in this sense is not temporal but indicates a requisite consideration of cause and effect to achieve a means–end logic. In fact, some practitioners use the *if–then* logical operand in place of *when*, such as "*If* function x, *then* function y," and test backward by asking "*Is* function y *caused by* function x?", but this added clarity has a time–cost penalty in terms of the duration available to build a FAST model.[6] All practitioners have to consider the external constraints that come as baggage with the modeling project. Both the *when* and *if–then* operands describe functional relationships that are closer to the mathematical use of the word *function* as x is mapped to y.

Referring to Figure 4.6, when you *Transmit Information*, you also *Store Information*. *Store Information* is an *independent support function* that supplements the dependent function *Transmit Information* with which it works. It does not fit comfortably to the right in the *how–why* path, so is not a dependent function. As an independent function it can be expanded in the *how–why* directions to build a subsystem FAST model but as an independent chain. Since the independent function is not on the major logic path, changing the function would not affect the basic function significantly. As this and subsequent chapters unfold, these terms will help you to glean insights and so will become very useful. The box located below the function *Transmit Information* is an activity. *Activities* are really types of processes or specific solutions that limit our range of alternatives but are sometimes necessary simply to get as good a FAST model as we can in the time available. That is, an activity describes a specific action that is initiated when the logic path function is activated. In Figure 4.6 it reads, "When you *Transmit Information*, you should *File Report*."

Since functions and activities can be described using a verb and a noun, a general rule to distinguish between the two is to look at both the verb and the noun. If the verb

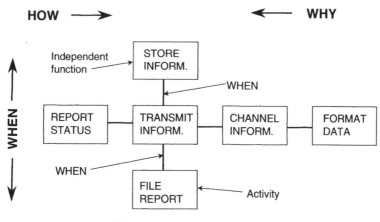

FIGURE 4.6 *When* Direction

describes a process activity such as store or hold, the intended action is really a solution that is creeping into our thinking. Alternatively, if the noun describes something specific such as a product, again it could be that the thinking is suggesting a preconceived way of doing things. Nouns that are generic are better to use when describing a function (in English grammar they are called *abstract nouns*). Independent functions (above the logic path) and activities (below the logic path) are the result of satisfying the *when* question asked in relation to the major logic path's *dependent function*. The acid test is whether the verb–noun opens the door to further innovation and the choice of alternative solutions, or locks us into a specific solution that reduces our chance of innovation.

KEY ELEMENTS OF A FAST MODEL

In order to learn, we need to standardize so that one person can pass on a method or insight to another, and this is also true of FAST modeling. At first some definitions may seem arbitrary, but they are simply part of a number of conventions that allow us to articulate our theories and check their reliability. As in mathematics, conventions enable one practitioner to examine and understand what another has written. Beginners are asked to learn these conventions so that they can communicate with other practitioners. Having said that, the next sections must be read so that the underlying theories and reasons for such conventions are understood. It is the lessons from the historical context that make the following conventions much richer than algorithms. Tools and techniques are only useful in the hands of someone who has mastered the theories of how to use them.

Before beginning the process of building a FAST model, it is important to understand some of the key elements of the model, shown in Figure 4.7. These elements and their definitions are described below.

Scope Lines

Scope lines represent the boundaries of a study and are shown as two vertical lines on the FAST model. How to determine the location of a scope line borrows from the

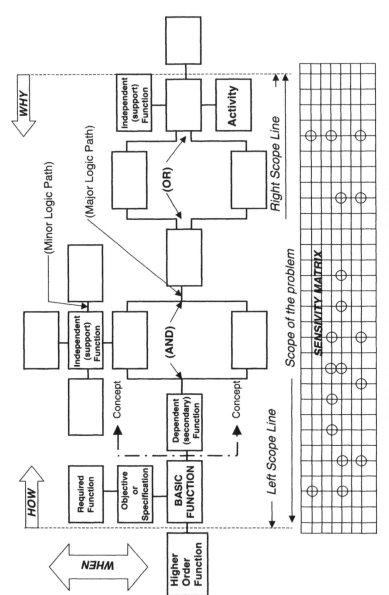

FIGURE 4.7 Graphic Layout of a FAST Model

68

discussions in Chapter 2. Scope lines bound the "scope of the study," that aspect of the problem with which the study team is concerned. Their existence is to encourage us to make our focus explicit, but we must not assume that they are there in reality. Our products function within other functioning systems. Scope lines are a simple and useful device to align a FAST model to usefulness, as they bound the study area, and contribute to the target evaluator, such as an external or internal customer.

The left scope line determines the basic function(s) of the study. The basic functions will always be the first function(s) to the immediate right of the left scope line. As the hierarchical means–end process suggests, this is what all the other functions inside the scope lines help to achieve. The right scope line identifies the abstraction level of the input function(s). Continuing to build the FAST model to the right of the right scope line will move the model to progressively lower levels of abstraction at the nuts and bolts end of thinking.

Highest-Order Function(s)

The objective, purpose, or aims of means–end logic, called the *highest-order function(s)*, are located to the left of the basic function(s) and outside the left scope line. Any function to the left of another function is a *higher-order function* because reading the FAST model in the *why* direction will lead you to the basic function(s) and the highest-order function of the subject under study, which allows a clear inference to the strategic value. In a means–end logic we are moving closer to ends, reasons for the usefulness created by the performance of functions on the *how–why* logic path.

Lowest-Order Function(s)

Functions to the right and outside the right-hand scope line, known as *lowest-order functions*, represent the input side and turn on or initiate the functions within the focus of study. Any function to the right of another function is a lower-order function and represents a means that is needed to carry out the function being addressed.

The terms *higher-* and *lower-order functions* should not be interpreted in terms of relative importance but as the external left- and right-hand sides of a functional explanation in which we are interested. As an example, *Receiving Objectives* could be a lowest-order function, with *Satisfy Objectives* being the highest-order function. How to accomplish the *Satisfy Objectives* (i.e., the highest-order function) is the scope of the problem under study. All functions on the *how–why* logic path should be "necessary".

Basic Function(s)

The function(s) to the immediate right of the left scope line, which represent the purpose or mission of the product or process under study, are called *basic function(s)*, as discussed previously. It is what a class of object would do; for example, all television sets perform essentially the same basic function and that's why customers consider buying a set before deciding on a particular model. The basic function is about a concept of usefulness in a relevant way. For example, "If I have no need to get from A to B, transportation functions have no value to me, for they are not useful at this particular moment in time." If the basic function is not valued by customers, all the functions supporting the basic function become unnecessary and the solutions that perform them become worthless or valueless.

Content

All functions to the right of the basic function(s) portray the intentional approach to enable the basic function to be performed. They are the means to a desired end. The functional explanation described by the linked functions should achieve the basic function(s). The content of a FAST model can represent either the current conditions ("as is") or proposed approach ("should be") of how a system should work. Which approach to use (current or proposed) in creating the FAST model is determined by the stakeholders, the task team, and the definition of the problem under study. Conceptually, all functions to the right of the basic function are treated as secondary functions, and because the *how–why* relationships have been joined intentionally, they can be changed. That is, if we map out the functions of method 1, then change method 1 to create method 2, some or many of the secondary functions would change.[7] Remember the NASA space pen and the Russian pencil from our discussion of elegant engineering solutions?

Requirements or Specifications

Requirements or specifications are conditions that describe the operating environment of the product or process; they are extrinsic. These parameters, specifications, or constraints, located above the *basic function*, must be achieved to satisfy the highest-order function of the system or product performing in its normal operations. Think of them as customer requirements. This logic comes out of a marketing paradigm and provides a means to allow customers and key stakeholders to use FAST models to direct subsequent creativity stages. Although the requirements are not intrinsic functions, they influence the concept selected to best achieve the basic function(s) and satisfy the end users' expectations. As we stated earlier, the alternative approach to modeling functions in VE is known as *customer FAST*.[8] This approach then uses a standard template to tease out the extrinsic functions that are used to represent what a design team has to achieve from a customer's perspective. Some practitioners favor including the requirements in the FAST model prior to the speculative or creative phase and argue that it channels the brainstorming process to come up with more practical ideas.

The channeling of ideas is the very reason that other practitioners choose not to include the requirements in the FAST model. They argue that including requirements limits the creative thinking process to preconceived solutions and discourages the consideration of new concepts for achieving the functions. The decision to include requirements in the FAST model and when to display them is a judgment call that depends on the problem definition and the objectives of the study. Including the requirements or specifications in the FAST model is optional.

Dependent Functions

As discussed previously, starting with the first function to the right of the basic function, each successive function depends for its existence on the one to its immediate left. That dependency becomes more evident when the *how* question and direction are followed. If any function on a *how–why* path is changed, it will affect all the functions to the right of the changed function (see Figure 4.5). Therefore, those functions affected are *dependent functions* because their existence depends on the functions to their left. Also, dependent functions located on the *how–why* logic path depend on the functions vertically above or below them in the *when* or *if–then* direction. We have to deal with causality in the *when* direction. For example, a gas engine doing what we want it to do also gets hot and needs cause–effect cooling solutions.

Independent (Support) Functions

Independent functions describe an enhancement or control of a dependent function located on the logic path. They do not necessarily depend on another function or method selected to perform that function in the use of *when*.[9] *Independent functions* are located vertically above the *how–why* logic path's horizontal *function(s)* and are considered secondary with respect to the scope, nature, and level of the problem being considered.

Logic Path Functions

Any function on the *how* or *why* logic path is a logic path function. If the functions along the *why* are chain-linked directly to the basic function(s) in a horizontal line, they are on the major logic path. If the *why* path does not lead directly to the *basic function* but runs parallel to the major logic path, it is a minor logic path (see Figure 4.7). Changing a function on the major logic path will alter or destroy the way the basic function was performed previously. If we want breakthrough innovation, this is where we aim. Changing a function on a minor logic path will disturb an independent (supporting) function that previously enhanced the basic function. If we want tame innovation, we examine the minor logic paths with a view to innovating there. Independent (supporting) functions exist to achieve the performance levels demanded by the implementation of solutions applied to the basic functions or because a particular approach was chosen to implement the basic function(s). For example, the power needed in a car's engine (i.e., independent functions) must be logically linked to the conditions required by the basic function and is why a Formula 1 racing car has a different engine than that in a family sedan.

ARTICULATING THEORIES IN FAST

Most examples of FAST models show variations in symbols and notations to help read the FAST model. Under current practice, the task team working the exercise creates its own language codes to aid in communications between team members. However, it is difficult for outsiders, even those familiar with the FAST modeling process, to read a FAST model without the aid of a participating team member. The need for a uniform graphic language becomes apparent when using a FAST model as a communication tool outside the team that created it. There is a balance to be achieved between a systematic way to communicate clearly that requires many conventions and the need to make a FAST model easy to read and accessible to people meeting it for the first time. Indeed, the trade-off between using the *when* or *if–then* operand is about how willing a team is to invest the time to think through very clearly or whether a less rigorous approach will suffice, given time constraints.

The following notations are offered as a recommendation for standardizing the symbols and graphical language of FAST. There have been a number of alternatives in

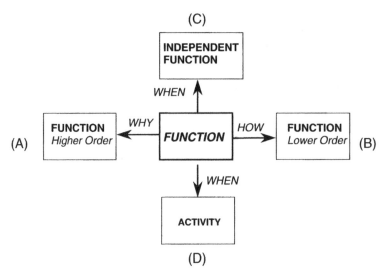

FIGURE 4.8 Four Primary Directions in a FAST Model

trying to adopt a convention that could be quickly understood, yet permit graphical flexibility to expand the model when needed. Why certain symbols were adopted will be explained where appropriate. However, the importance is in creating a common language that accurately captures the intended communications, not in debating what or where a symbol is used. If FAST models are to be universally accepted as a powerful problem-structuring and team-building process, they must first be universally understood.

The four primary directions in a FAST model are shown in Figure 4.8. The *how* and *why* directions are always along the logic path, whether it is a major or a minor logic path. The *when* direction indicates an independent or supporting function (above) or activity (below).

An important rule to follow when addressing a function is always to read from the exit direction of a function to question the three primary questions *how, why*, and *when*. The questions are answered in the corresponding exit block, as indicated below.

- *How* is [Function] to be accomplished? By performing B.
- *Why* is it necessary to [Function]? To perform A.
- *When* [Function] occurs, what else happens? C or D.

VARIATIONS OF *HOW–WHY* QUESTIONS

It may sometimes be necessary to modify *how–why* questions to fit a specific study issue. Some variations in asking *how* questions are:

- How is [Function] proposed to be accomplished?
- How is [Function] actually accomplished?
- How would I [Function]?

Some variations in the *why* direction are:

- Why is it necessary to [Function]?
- Why should you [Function]?
- Why would I [Function]?

All of the foregoing questions can be used in FAST modeling, but it is best to stay with the *how–why* questions selected initially for resolving a particular problem.

Answers to the *how–why–when* questions can yield multiple options, so we require logical operands such as *and* or *Or*. Also, the relative importance of functional relationships can be suggested as either equal or unequal in terms of priority; we explain this further in the next section.

Considering *And–Or* Along the Logic Path

In many instances, the answer to the *how* or *why* questions require more than one functional response. The answer could be in the form of *and* or *or*. The logical operand *and is* represented by a fork, whereas *or* is represented by a clear and unambiguous split in the logic path. Figure 4.9 shows two examples. The one on the left shows *and*, which reads "How do you *Confirm Compliance*?", which is answered by "*Verify Documentation 'and' Validate Performance*." On the right of Figure 4.9 is an *or* relationship for an electrical switch that reads "How do we *Manage Current*?" and yields an answer "By either *Complete Circuit 'or' Interrupt Circuit*."

In Figure 4.10 we show how equally important functions are balanced symmetrically with respect to the *how–why* logic path. This reads "How do you *Confirm Compliance*?" "By *Verify Documentation 'and' Validate Performance*." In Figure 4.11 we can see the question "How do you *Isolate Fault*?" and the answer "By *Simulate Environment 'and' Analyze Complaint*." However, *Analyze Complaint* sits below the major logic path like a hammock, which shows a different *and* relationship to that above. In Figure 4.10 the split is drawn evenly to show that *Verify Documentation* and *Validate Performance* are equally important. In Figure 4.11, *Simulate Environment* is shown as more important than *Analyze Complaint*.

In example A in Figure 4.12, showing a planned maintenance FAST model, the answer to the question "How do we *Maintain Operation*?" is answered "By *Repair*

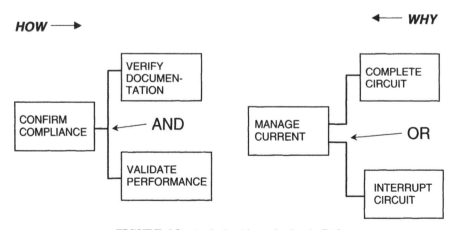

FIGURE 4.9 *And–Or* Along the Logic Path

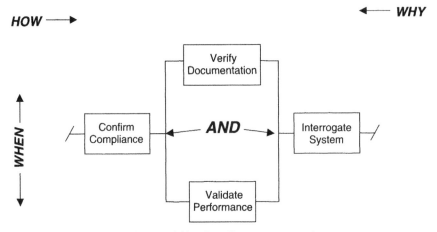

FIGURE 4.10 Equally Important *And*

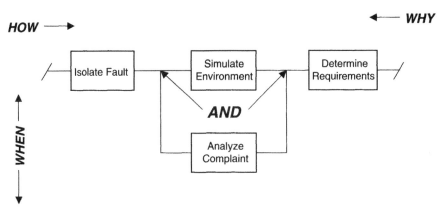

FIGURE 4.11 Less Important *And*

EXAMPLE A
"OR" (Equally Important)

EXAMPLE B
"OR" (Less Important)

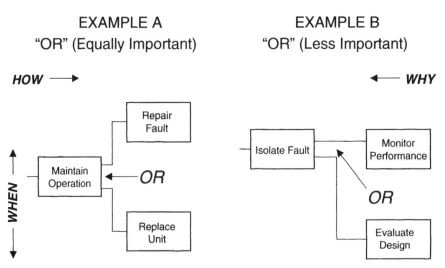

FIGURE 4.12 *Or* of Equal and Less Importance

Fault 'or' 'Replace Unit'"; we either mend a broken part or we replace it. This is a clear choice. As we model the *or* operand, always try to represent it as an *exclusive-or*. *Inclusive-or*, such as "You can have apples and custard '*or*' just apples '*or*' just custard" introduces an ambiguity that might cause us to miss articulating our theories clearly. In example A, both functions are equally important, but only one is used. Which function path to select depends on the value that each path enables in a particular situation; sometimes it might be best to repair, and sometimes it might be best to replace. Reading in the *why* direction, one path at a time is followed. Therefore, "Why do you *Repair Faults?*" is answered by "So that you can *Maintain Operation.*" Also, "Why do you *Replace Unit?*" "So that you can *Maintain Operation.*"

The same thought process applies to example B in Figure 4.12, taken from the analysis of a quality assurance department, except that *Evaluate Design* is shown as being less important than *Monitor Performance* in order to *Isolate Faults*. As in the *or* example (example A), only one path is selected, but the preferred path is shown graphically as *Monitor Performance*.

Considering *And* in the *When* Direction

Multiple *when* functions (applicable to independent functions and activities) that are read as *and* are shown in Figure 4.13 by connecting boxes stacked above or below the logic path functions. The segment of a sales department's FAST model states: "*When* you *Influence Customer*, you *Inform Customer 'and' Apply Skills.*" If it is necessary to rank or prioritize the *and* functions, those closest to the *how–why* logic path function should be causally more influential. Stacking the functions in the *when* direction (above or below the function) to indicate *and* allows those *and* functions to be expanded in the *how* and *why* directions, if necessary, to open the way to create minor logic paths.

Considering *Or* in the *When* Direction

Occasionally (but not frequently) we come across a situation where *when* has a number of *or* functions stemming from a particular function on the major logic path and offering choices in the *when* path. *Or* functions are not stacked but configured as *flags* to indicate optional paths. Locating the *flags* to the left or right of the vertical line affects how you read back into the logic path function. It is important to select which

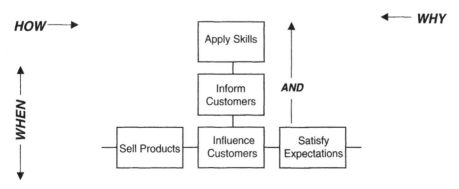

FIGURE 4.13 *And* Along the *When* Path

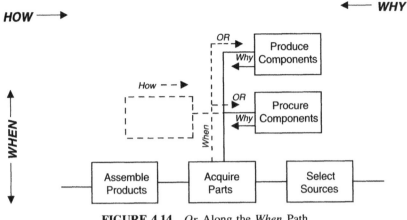

FIGURE 4.14   *Or* Along the *When* Path

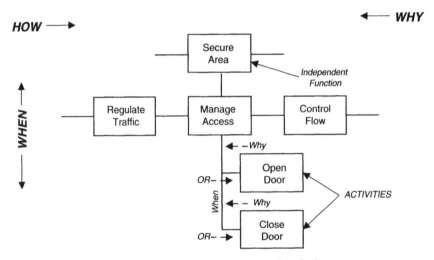

FIGURE 4.15   *Or* Along the *When* Path

direction the flags face, because the exit line of the function will address *why* or *how* those functions are performed.

Figure 4.14, taken from a manufacturing facility FAST model, asks "How do you *Assemble Product*?" "By *Acquiring Parts. When* you *Acquire Parts*, you *Produce Components 'or' Procure Components* (not both)." Reading from the *or* functions, "Why do you *Procure Components 'or' Product Components*?" "To *Acquire Parts*." This is logically coherent and so makes sense. If the *or* functions were facing the other way, reading from the *or* path would ask "*How* do you produce or procure components?" which would be answered "By *Acquire Parts*." This does not make sense because you don't produce or purchase components by acquiring parts. Remember, the validation process requires the logic to stack up in both the *how* and *why* directions, so care should be taken in determining which way to position the *or* functions.

Figure 4.15 shows a part of a traffic study in a plant layout. Note that the *or* flags *Open Door* and *Close Door* are activities. The same rules apply to activities as apply to functions. In this example, addressing the function *Manage Access*, as we ask "What occurs *when* you *Manage Access*?", the answer, reading down, is "*Open Door 'or' Close Door*," which are activities. Reading above, the answer to "What occurs *when* you *Manage Access*" is *Secure Area*, which is an *independent function* of *Manage Access*. Reading the activities back, the functions face the *why* direction and the question reads, "Why do you *Open Door 'or' Close Door*?" To *Manage Access*.

FAST MODEL-BUILDING PROCESS: PRODUCT EXAMPLE

Understanding the FAST modeling process up to this point will cover most of the applications that a practitioner will meet in the context of a five-day VE workshop. Before proceeding with other graphical notations to complete the FAST language, we will apply the lessons learned in creating a FAST model of a simple hardware product.

The project selected is a disposable cigarette lighter (discussed in Chapter 2), shown in an exploded view in Figure 4.16. If asked what the basic function of a cigarette lighter is, at this point in the conversation the most probable answer would be to *Produce Flame* or *Light Cigarette*. However, the FAST modeling process will demonstrate that it is best to reserve judgment in selecting the basic function(s) until the FAST model is structured.

There are 19 components in this product, including the fuel. If we were using a detailed function analysis such as the random function determination (RFD) (see Chapter 3), we would take the 19 components, determine each component's functions, and then select the component's basic and secondary functions. We would then try to expand the intentional logic in a *how–why* direction out of the assumed basic function to start the FAST model-building process.[10] To keep the focus on the FAST process rather than on the 19 detailed parts, we have selected a combination of six

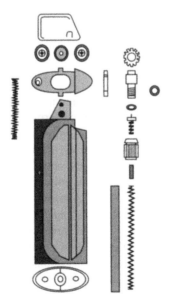

FIGURE 4.16   Disposable Cigarette Lighter

TABLE 4.1   Random Function Determination

Component	Function	B	S
Cigarette lighter	Produce flame		
Body	Contain fuel Support components Accommodate hand		
Valve assembly	Release fuel Control flow		
Wheel assembly	Strike flint		
Flint	Produce spark Energize particles		
Shield	Protect flame		

subassemblies and components and displayed them in random function determination (RFD) format, to start the FAST model (see Table 4.1). What is important to note is that the components provide a link between our thoughts and reality that keeps us grounded. It is as if we are using the product as a gateway to the original designer's thinking of how the lighter was created. The object of this part of the process is to form a reference point for subsequent analysis. Note the blank columns in Table 4.1 headed "B" and "S" for "basic function" and "secondary function." We did not go through the process of determining which functions are basic or secondary, because with only 10 functions identified, we elected to use all of the functions noted for the next step in constructing a FAST model.

Expanding the Number of Functions

Place the functions selected in the center column as shown in Table 4.2 and ask of each of the functions, "How is that function performed?" The answer, in the form of another function, is placed in the right-hand column under *how*. Then ask *why* that function is performed and place the answer in the left-hand column under *why*. To test the logic of the response, read across the row. Note that the function *Energize Particles* in Table 4.1 was placed in the *how* column in Table 4.2 because it directly answered the question "How do you *Produce Spark*?" In Table 4.2, the response to "How do you *Produce Flame*?" is *Ignite Fuel*, and the answer to "Why do you *Produce Flame*?" is *Produce Heat*. To test the logic of the answers, read across the row. In the *how* direction (note the arrow direction), "How do you *Produce Heat*?" is answered By "*Produce Flame* and "How do you *Produce Flame*?" is answered" By *Ignite Fuel*. Reading in the *why* direction, "Why do you *Ignite Fuel*?" begets "To *Produce Flame*" and "Why do you *Produce Flame*?" begets "To *Produce Heat*." If the knowledgeable team is comfortable with the answers, the explanation is a good one and the logic has

TABLE 4.2   Expanded List of Functions

←———Why	Function	How———→
Produce heat	Produce flame[a]	Ignite fuel
Add O_2	Release fuel[a]	Open valve
Control flow		Depress level
Create friction	Strike flint[a]	Rotate wheel
Ignite fuel	Produce spark[a]	Energize particles
Transport fuel	Contain fuel[a]	Enclose fuel
Operate lighter	Support components[a]	Assemble parts
Control motion	Accommodate hand[a]	Shape container
Manage flame	Control flow[a]	Restrict exit
Ignite tobacco	Protect flame[a]	Control environment

[a]Function taken from the theory of random function determination.

been satisfied.[11] Note that the location of the *how* and *why* words are reversed, but the arrows are properly placed so that the *how* (read right) and *why* (read left) answers are properly oriented. This places the *how* and *why* responses in the correct columns while maintaining the direction of the logic questions.

Case for Using Active Verbs

The reason for using active verbs becomes apparent when asking the intuitive *how – why* questions, the questions we ask of the verb. *Ignite, Produce, Control*, and so on, are active verbs that should be answered with another active verb. Active verbs are needed to make the FAST model a stand-alone communication tool that documents how a system works. Using passive verbs such as *Provide, Meet, Review*, and so on, may satisfy the immediate requirements of the team but would be not communicate the intention of the function to others outside the team without defining or verbalizing the function further. For example, what would the function *Provide Satisfaction* mean to you if your job were to make sure that it was performed adequately? If *Provide Satisfaction* were replaced with *Reduce Complaints*, you would have a better grasp of the actions to take.

Purpose of Expanding Functions

The purpose of expanding the number of functions in this manner (see Table 4.2) is to generate active functions that relate to the study topic and will later form a chain of logically dependent links when building the FAST model. See this as a starter kit that will make the building of a FAST model easier. This does not mean that you must place the functions in sets of three in the FAST model. On the contrary, after functions have been transcribed on small Post-It notes, the sets should be broken up and treated as individual functions so that the logic of the means–end of the FAST model, referenced to strategic aims, drives the selection process.

Avoiding Duplicate Functions

Try not to repeat functions when generating them and avoid using different words for the same thing. The more precise the function definitions, the better the FAST model. If functions are duplicated, such as *Ignite Fuel* in the example, the duplicates are omitted when transcribing the functions on the Post-It notes. Make sure that the nouns and the level of abstraction selected (see Chapter 2) are consistent, for if we have the two functions *Ignite Fuel* and *Burn Butane*, we are again bringing unnecessary ambiguity to our thinking and end up playing semantics. Also, a center column function may better satisfy a *how–why* question, as shown in Table 4.2, where *Energize Particles* best satisfies the *how* question of *Produce Spark*.

Starter Kit Functions

In all probability, every function generated in the expansion process will not be used, just as the list of functions produced may not include some functions needed for the FAST model. One of the purposes of the function expansion process is to create a starter kit of related functions to begin the FAST modeling process. As a starter kit the functions used in building the FAST model will be tested again to determine if the FAST model is valid and fits with the knowledge of the team's view of how the system works.[12]

PREPARATIONS FOR BUILDING A FAST MODEL

The supplies needed for building a FAST model include a large sheet taken from a roll of butcher paper or flipchart paper taped together, Post-It notes, soft lead pencils, and a large, soft eraser. After placing the butcher paper on the wall, visible to the full team, prominently mark the *how–why–when* directions on the top corners of the paper. The Post-It notes should be $1 \times 1\frac{1}{2}$ inches. Transcribe each function on a separate Post-It drawn from each of the three columns (eliminate duplicates), and ignore the column headings. The functions will be treated individually and are placed on a separate sheet of paper next to the rolled-out butcher paper. Consider the individual function notes as puzzle parts, and the assignment is to build the puzzle using the function logic.

BUILD *HOW* AND TEST *WHY*

The most difficult part of building a FAST model is placing the first Post-It note with the individual verb–noun function onto the butcher paper. It looks very lonely. Don't be concerned whether it was the right one to choose or if it felt like it could be the basic function. Although a FAST model can be developed by selecting any function at random and asking the *how* and *why* questions, it is better if you select a function you think could be a basic or higher-order function so as not to lose sight of the usefulness that a basic function has with respect to customer value.[13] Following this first function, we ask "How do you achieve this function?" The answer, "By ...," should be found as another function and placed to the right of the first function. You now have two functions that are expanded to become three and then four, and so on.

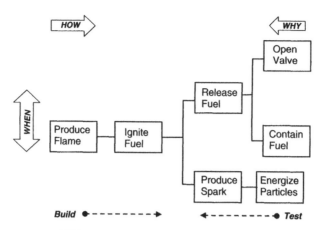

FIGURE 4.17　Building the FAST Model

Using this process, we will build the FAST model in the *how* direction and test it in the *why* direction.

In the cigarette lighter example (see Figure 4.17), we can start with the assumed basic function *Produce Flame*. Since we are constructing an "as is" model, the answer to the question "How do you *Produce Flame?*" (see Figure 4.17) would be by *Ignite Fuel* and "How do you *Ignite Fuel?*" by *Release Fuel 'and' Produce Spark* (note the *and* split). Select functions from among those already transcribed on the Post-Its. If a function seems to be missing because the logic does not flow, create a new function on another Post-It and position it on the evolving FAST model. Continue the model-building process in the *how* direction following the split *and* paths (see Figure 4.17). "How do you *Release Fuel?*" The answer is another split with the *and* path, *Open Valve 'and' Contain Fuel*. Following the lower branch we ask, "How do you *Produce Spark?*" to get the answer, *Energize Particles*.

Relationship of the Left Scope Line to the Basic Function

When you feel that you have reached the end of the *how* path, go back to the starting function *Produce Flame* and test to the left (see Figure 4.18). Extending the FAST model in the *why* direction will enable you to test the function selected as basic to ensure its adequacy. In the cigarette lighter example, ask, "Why do you *Produce Flame?*", in order to *Create Heat*, and "Why do you *Create Heat?*", in order to *Ignite Tobacco*. When you have completed, tested, and satisfied the major logic path, the remaining functions and activities can then be placed on the model to satisfy the *when* question. Have you identified the basic function in the cigarette lighter example?

The scope lines in the partially completed FAST model illustrated in Figure 4.18 have been left off so that you can determine where you want to limit or scope the problem and make your focus explicit. After extending the logic path in the *how* and *why* directions, it must be decided where to place the scope lines, which also bound the focus of the problem under study. A question at this point may be "When do you stop building the FAST model, and how do you locate the scope lines?" The answer resides in the problem definition that initiated the FAST model. Returning to Chapter 3 and reviewing the section titled "Rules Governing Basic Functions" would be helpful in deciding the basic function. We reveal our answer later in the section "What's the Problem?".

Determine the Basic Function – Place the LEFT Scope Lines

HOW →

← WHEN →

WHY →

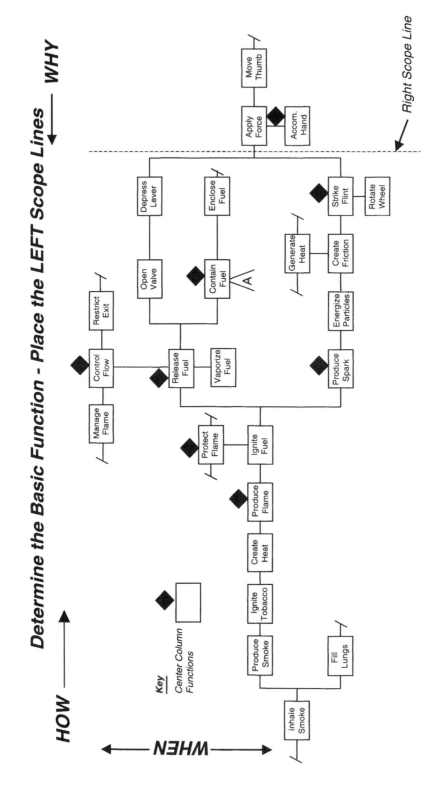

FIGURE 4.18 Seeking the Basic Function

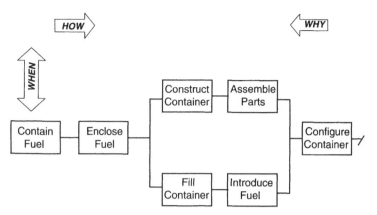

FIGURE 4.19   Lower Level of Abstraction

Right Scope Line

On the right side of the FAST model (*how* direction), we probably aren't concerned about the physical functions of how a thumb is activated, except that it is used to rotate the wheel (see Figure 4.18). The input (right) scope line would probably be located to the left of *Apply Force*.

Left Scope Line

To help us to locate the *left scope line*, we must remember that the purpose of the value methodology, in which FAST modeling is so central, is to increase value. This is done through better performance of real-world solutions that we coordinate and optimize through the concept of functions. We have to make the conceptual link between why we innovate and what a FAST model corresponds to. Note the symbol under the function *Contain Fuel* in Figure 4.18. The "A" under the inverted "V" signifies that the function *Contain Fuel* has been selected to be expanded for a more detailed analysis, as shown in Figure 4.19. Starting at a higher level of abstraction allows for an uncluttered macroanalysis of the overall problem until those key functions can be found, isolated, and the problem resolved. Here we have two foci, but for clarity they are modeled as two separate FAST models.

WHAT'S THE PROBLEM?

Prior to starting the FAST model process, the problem, or opportunity issues to be resolved, or outcome to be achieved by the model, must be understood. This is essential because the model will reflect the strategic focus caused by the initial problem definition. The definition triggers the need to model a system to diagnose what needs to be done to find an elegant solution. In the lighter example, let's assume that the problem involves a manufacturing company looking for opportunities to reduce their product cost as a way to maintain profits and sales after matching competitive price pressures. Let's assume further that the same company produces butane fuel for this product and that butane fuel accounts for a major part of the company's business. Under these conditions, where would you place the left scope line?

If you place the scope line between *Ignite Tobacco* and *Create Heat* (see Figure 4.18), you have chosen to designate *Create Heat* as the basic function. But is your choice correct? By definition, the location of the left scope line designates the basic function and we've said earlier in this book that the basic function is about usefulness, and its from concepts of *useful* that pragmatic views of *value* are to be found. If we satisfy the basic function by using a flameless source of heat such as an automobile cigarette lighter, how would that affect the company's butane fuel business? Under these conditions, *Produce Flame* would probably be selected as basic, as it provides the means to infer links to strategic outcomes. Conversely, if the study involved the development of a new concept for igniting tobacco products, *Create Heat* would be considered basic. So the left-hand scope line would be placed between *Create Heat* and *Ignite Tobacco because* of the need for innovation to find better ways to *Create Heat*.

To further emphasize the importance of the role that problem definition plays in FAST modeling, consider the following scenario: If a diminishing sales and profit problem were traced to a breakdown in the distribution system and a massive buildup of inventory, studying the product's functions in a FAST model would not resolve the root cause. The problem is not in the product but in the business process.[14] While FAST modeling is a very effective method for understanding complex problems and helping interdisciplinary teams gain insights that lead to focused innovation, it is essential to ensure that the focus of the FAST model addresses the <u>right</u> problem.[15] Remember our analogy of a doctor and the diagnosis phase. What value is a painkiller to a patient when surgery is needed? We must always conceptually link the product or project to the organization in which it exists in the same way that a doctor considers a particular ailment or malfunction with respect to a patient's well-being. Systems exist within systems, and they operate through the performance of their functionality.

Defining the Problem

Problem definition is not only important to focus the FAST model on the key problem issues but also serves to determine which disciplines make up the FAST modeling team. The value of the FAST model flows from its proximity to a true and accurate representation of reality. This requires experts with appropriate knowledge to help represent how a system really works in terms of its functions. The structure of the FAST model can be used as a guide to choosing the disciplines for the team's composition. In selecting a team to resolve a business process reengineering issue, the scope lines partition three areas, within the scope lines and on either side of them, from which we consider how to select the team. Participants from the right (input) side of the right scope lines would represent those who own the problem or are sponsors of the problem resolution. Those selected from the area between the scope lines would represent those responsible for the resolution of the problem (e.g., management and engineering disciplines). Participants selected from the left (output) side of the left scope line represent those functions that are affected by the corrective actions that would resolve the problem (e.g., subcontractors). As an example, in restructuring a product engineering department using the FAST process, consideration should be given to assigning representatives from production and sales as the "output" side of the team; marketing, finance, and business planning personnel could represent the "input" side; and the product engineering staff could represent the scope of the problem, which is between both scope lines.

Three Questions Before Starting the FAST Process

Before the team starts to develop the FAST model, it is essential to agree on the problem statement, and this consensus must be linked to the supporting data. When we state that "*xyz* is a problem," we reveal our theories, which should be tested against data and evidence to ensure that our founding premises are valid. Those senior managers who will subsequently provide funds to implement solutions flowing from the FAST model must also champion and direct the process. A common objective, a direction, and the level of abstraction must be decided to keep the focus on the core issues rather than peripheral issues (i.e., to avoid treating symptoms). If developed properly, contained within the problem statements are the higher-order and basic functions of the problem. These can be identified by underlining the active verbs and nouns in the problem description and later used to start the FAST process. Some practitioners use system dynamic models[16] to understand what forces are at work and which outcomes need closer attention. A system dynamics model can run simulations to show how various problems and opportunities might present themselves if trends run as they might. It is this type of strategic enquiry that is the starting point for both VE and FAST modeling. Once the strategic context, in which the study will take place, has been understood and grounded in the evidence, we can ask three fundamental questions to characterize and frame the problems and opportunities we are to address.

* *Question 1*: What problem or opportunity are we about to resolve?

The answer to this question should be expressed in no more than one small paragraph or a few simple statements. The root cause is usually not complex, but it is hidden by its symptoms or imposed solutions. As such, cost reduction, normally stated as a problem, is not a problem; it is a potential solution to the root cause. The team should be made aware of the issues concerning management and why they feel the problem will be resolved by cost reduction. This is not to say that we are against cost reduction but that we see it as a potential solution whose relevance flows from the consequences it would lead to. Sometimes, organizations need to be subjected to cost reduction studies in the same way that a fruit tree needs occasional pruning to ensure its full potential. This question is asked within a strategic enquiry phase. It provides the team with an opportunity to explore other options that could resolve the fundamental issues in more effective ways.

* *Question 2*: Why do you consider this a problem or opportunity?

The answer to this question is where the symptoms of the problem show up. Two purposes are served by asking this question. First, the underlying causes are explored while discussing perceived answers. Second, the *why* in the question is also directing conversations toward higher-order functions. Therefore, by isolating function terms from the answers to this second question, we can match them to functions in the answers to the first question to determine if there are any function dependency links. In other words, we quickly build a generic FAST model in our minds to see if we can identify a cause of disvalue and the knowledge needed to address it. It is this cause that points us to the areas we should focus on, and by writing things down we make them explicit.

- *Question 3*: Why is a solution necessary?

The answer to this question should support and/or modify the answer to the second question and identify additional functions to explore. Another version of this question is: "What are the consequences of not solving the problem?" This is where we test our assumptions as to whether the team's time and energies are being put to the most valuable use. Is the problem worth solving?

The reality of the three questions above is that we often spend several hours helping managers to structure the problem or opportunity that they want someone to resolve. If the problem concerns the restructuring of an organization, the three questions should be supplemented with descriptive statements such as the mission, charter, and goals of the organization. Additional functions can be selected from these documents and placed in the center column of the function expansion process discussed Chapter 3 in relation to random function determination.

How the Strategic Questions Are Asked in a Workshop

When asked the first question, the problem owner is apt to answer the second question. As an example, let's say that the response to the first question, "What problem or opportunity are we about to resolve?" is "Our profit margins continue to decline and can no longer support our operating expenses, so we must cut costs." Then we hold this response against the second question, "Why do you consider this a problem or opportunity?", so that we can unpick the network of considerations that led management to such a view. We unravel the theories that abound and make complexity more prevalent. After placing the response to the first question under the second question, we can now ask the problem owner, "What conditions exist that have created the problem you describe?" That response, which seeks out the causes of the previous reply, will be a better answer for the first question.[17] We are trying to delve below our superficial inferences to understand the root causes. These underlying causes are the dynamic results of different systems vying for domination, and the superficial thinker fails to delve deeply enough to resolve them. As an example, the response to the question above could be, "More competitors have entered the market, making our product price sensitive." Then the problem that should be addressed is, "How can we improve our competitive position and protect our profit margins" rather than limiting ourselves to preconceived solutions, such as "Cost cutting is what we really need."[18] Cost reduction would be one among many options to improve the competitiveness of the company's products.

SYMBOLS AND NOTATIONS USED IN FAST MODELING

The use of five words governs how a FAST model is constructed, read, and analyzed. The words are also supported by standard graphic symbols, which enable the transfer of information outside the team that created the model. Figure 4.20 identifies the words and symbols used in FAST modeling. Some of the more common notations and symbols used to express events or highlight dimensions on the FAST model are shown in Figure 4.21.

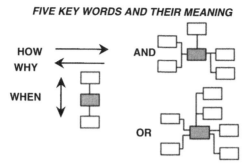

FIGURE 4.20 Reading FAST Models

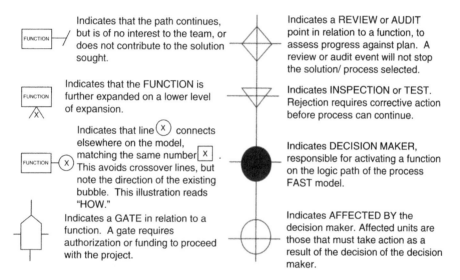

FIGURE 4.21 FAST Model Symbols and Notations

TAKING EXCEPTION TO THE FAST RULES

Some liberties in the procedural rules are acceptable, with the exception of the logic process expressed in the *how–why* and *when* directions. This is to facilitate the practicalities of building a FAST model in a time-constrained workshop. However, it is suggested that the principles of FAST modeling be well understood before entertaining a change to the rules for convenience sake and running the risk of degrading the quality of the model and the explanation it represents. Some conditions that may call for *bending the rules* are discussed next.

Independent Functions Above the Logic Path, Activities Below the Logic Path

On large, complex models it is often necessary to place some independent support functions below the logic path and activities above the logic path to achieve some degree of graphical balance. Many independent functions are expanded to create minor logic paths, where most activities are listed in the stacked *when* direction. If you can keep to the convention, only activities below will help you to see the opportunities for process change solutions more clearly. If the *when* operand is substituted for the *if–then* operand, this flexibility is lost, as reducing ambiguity becomes the modeler's main purpose. This flexibility cannot be accommodated with the *if–then* rule, though. The terms *when* and *if–then* can usually be interchanged. Both are saying, "The performance of this (*x*) function causes (*y*) function to occur." Or, reading back, "(*y*) function is a direct result of (*x*) function happening." Treating *when* as a time dimension and not as a way to express cause and effect produces ambiguity.

No Activities in the Major Logic Path

When creating a FAST model it is best to focus the team's attention on satisfying the *how–why* question even if some activities below or above the logic line are used to satisfy the logic flow. After completing the logic path, to raise the FAST model's level of abstraction, activities can be removed and placed under the appropriate function, provided that the logical flow still satisfies the *how–why* questions. What must be realized is that an activity is really a process grounded in a day-to-day way of doing something. We use it here as a quasi-function but argue that we should probe deeper to define the proper function. For example, for convenience we might have a function *Store Product* in a major logic path between *Create Product* and *Distribute Product*. It is only when we ask why we *Create Product* in this consideration that we realize that we do not make things to store them but to sell them, so both *Store Product* and *Distribute Product* are activities born out of how we currently do business. They are not really functions, even though they have the verb–noun construct, because their existence is not really valuable to the system in which they reside; functions provide a logically vital necessity to those to the left of them in the *why* direction. To keep this explanation brief, let us explain that *when* we *Store Product*, we do so to balance the output of one system with the input of another, so a proper function would be along the lines of *Balance Systems*. This should not be on the major logic path if it is not a means to the desired end. *Distribute Product* is a proper function in the *when* direction because other functions, such as those related to where customers live, necessitate that a manufactured product has to be moved from the point of manufacture to the point of sale.

The reason that we don't always get the specific or activity functions out of the FAST model is because in the context of five-day VE workshops, the time constraint of a single day set aside to build a FAST model often prohibits deeper consideration. We simply have to live with the quality achievable and rely heavily on the practitioner's skills in such situations. However, we have worked on some projects where we have been allowed a number of weeks to get the FAST model much closer to an explanation that is an accurate reflection of reality. A better model then allows us to consider and select from many solutions to perform the functions in the best way. When the quality of the FAST is more important than the time it takes to build it, we use *if–then* rather than *when* and check that what the team says is the case can be corroborated by evidence from observation or experiment.

The process of removing activities from the major logic path raises the FAST model to a more abstract, macro, or higher level of abstraction and closer to a collection of proper functions that reflect the pure intentionality that is independent of any solution, method, or technique. Inserting activities in the major logic path lowers the FAST model to a micro or lower level of abstraction and locks us into a narrow range of options, as it becomes context or case specific. When adding dimensions to the FAST model (discussed in Chapter 5) it is best to raise the FAST model to its highest practical level of abstraction, to focus on satisfying those dimensions that will resolve the problem issues by widening the solution space.

Only Two Words Used to Describe Functions

Trying to find the two words that express a function adequately can be a frustrating experience and should not be abandoned too quickly. Sometimes a third word is needed to communicate and clarify the function. This usually occurs when the subject of the FAST model is highly technical. Some random examples noted are *Control Feed Temperature, Recycle Process Heat, Position Femoral Component*, and *Develop Test Plan*. Here *Feed Temperature*, a conjunct noun, and the other examples, which read like split infinitives, must be seen as a single concept being communicated through semantics as is intended with a single noun; no matter what semantics we use, our intention is to show an action operating on a thing which contributes benefits to the object or system in which it resides. As stated previously, we want an active verb to suggest an action operating on a thing described by a single noun. In other words, read *Feed-Temperature* as a single hyphenated word so that the single concept it refers to is the focus of the noun placeholder in the verb–noun label. That we have to describe the thing with more than one word is more to do with the syntax of the English language than our aim to talk about a single thing. Sometimes abbreviations or acronyms are used as verbs to satisfy the two-word rule, such as *Shield RFI* for *Shield Radio-Frequency Interference*. A danger in taking liberties with the two-word function rule is that it may lead to more words, then sentences and paragraphs, ending up not as a concise view of a function, but as prose. This should be avoided at all costs. A rule of thumb to remember is: The more words you need to express a function, the less understanding you have of its proper function.

LOOP-BACK MODELING

Occasionally, processes with a recurring phase will find it necessary to duplicate a depiction of this repetition. Ideally, the FAST model would not show such repetition because it is a product of how we implement solutions rather than focus on what function needs to be performed cyclically. The functionality of an elegant solution should not require the remodeling of the repetition, as we should search for a better technology to perform this reiteration once and once only. The practical reality is that often the "ideal" solution does not exist at this time, so we must recognize the innovation opportunity even if we don't know how to seize it. The principal reason for the need to loop back through a process phase is that something has changed or needs to be changed in relation to the performance of solutions. In value management,[19] loop-back actions are considered as non-value-adding activities, the reason being that if relevant information was available, the necessity for repeating a phase could have been avoided, reducing costs and time. Sometimes, especially in technical processes,

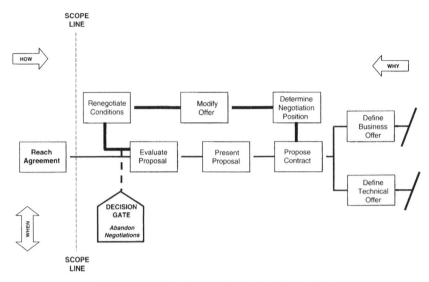

FIGURE 4.22   Contract Negotiation Loop Back

adaptive technology is lacking, which necessitates repeating the process to achieve the desired outcome.

Two examples are offered to describe and show how to display the loop back graphically on a FAST model. Figure 4.22 describes a loop back that commonly occurs during the negotiation process. In the contract negotiation FAST model in the figure, note the option path at the *why* exit of *Evaluate Proposal*. If the terms and conditions are acceptable, the path and answer to the question "Why *Evaluate Proposal*?" is "To *Reach Agreement*." However, if the parties cannot agree, the second path in the *or* gate is followed. In this latter case the answer to the question "Why *Evaluate Proposal*?" is "To *Renegotiate Conditions*."

Following that logic path, the answer to "How do you *Renegotiate Conditions*?" is *Modify Offer*. And how you *Modify Offer* is by *Determine Negotiation Position*. As that position is realized, you can then *Propose Contract*. In so doing the process is repeated, forming a loop back. The loop-back process is repeated until agreement is reached, or as the gate symbol implies, the negotiations are abandoned.

Validating the Logic Flow

Reading the FAST model in Figure 4.22 from left to right when asking "How do you *Evaluate Proposal*?, note that the *or* gate is not available in the *how* direction. Continuing in the *how* direction, we can ask "How do you *Present Proposal*?" "By "*Propose Contract*." *When* you *Propose Contract*, you cause *Determine Negotiation Position*, and "Why do you *Determine Negation Position*?" "To *Modify Offer*".

If the loop back reads logically in the *why* and *how* directions, the logic flow is validated. As the FAST model shows, if relevant information was available to the negotiating party's and they were of a single mind, before addressing the function *Propose Contract*, the loop-back process could have been avoided.

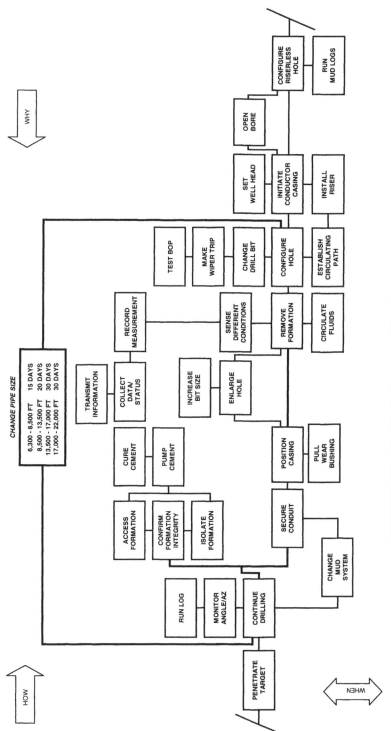

FIGURE 4.23 (Partial) Exploration Drilling Model: Loop Back

91

Exploration Drilling Model

Figure 4.23 is a technical example of an offshore exploration drilling project to determine if oil reservoirs warrant production drilling. A part of the complete FAST model relating to the loop-back condition is illustrated in Figure 4.23. This loop-back example shows when the approximate number of feet have been drilled, the pipe string is raised and resized, casings secured, and the drill bit changed prior to continuing drilling.[20] The estimated number of drilling days required for each loop-back phase is also shown as a FAST model dimension. The drilling depth determines the number of times that a pipe string is raised and resized. In this case technology to reach and penetrate the target in one drilling pass is not yet available.[21] By highlighting the existence of a loop back, the potential for researching the drilling process to determine other innovative ways of reducing or eliminating loop backs is exposed. That is, a loop back is an indication that an elegant engineering solution is not yet in common practice. The loop-back system is a simple way to illustrate graphically that a part of a process is repeated, or looped back, rather than extend the FAST model to show (in this case) the same process element repeated four times in the FAST model.

CLOSING REMARKS

Practitioners who have been involved in a number of FAST exercises will agree that the dialogue and discussions occurring in the process of model development, stimulated by the logic questions, are often more important than the model itself. The process of identifying functions, questioning, and justifying them, especially by an interdisciplinary team, is particularly useful in structuring a problem and moving toward corrective actions. Once the problem has been structured, the model serves mainly to explain the team's rationale to outsiders and to communicate across different disciplines. As important as the model is, it is a tool to stimulate creative thinking. It is not an end product unto itself, but a beginning. The end product is satisfying the issues that the FAST model has helped to define and resolve. In the Appendix we offer practical hints based on frequently asked questions that will help you to improve your modeling skills. In the next chapter we discuss how solutions tied to functions are measured and how such dimensioning allows us to analyze what the FAST model is revealing.

<hr>

CHAPTER 5

DIMENSIONING THE FAST MODEL

Looking at a FAST model, one can appreciate the power of the process in helping teams to understand how things work. However, to bring out its full potential requires building a FAST model to a case-specific project with clear performance measurements. To this end, dimensioning the FAST model brings FAST to a higher level of understanding. It not only acts as a stimulus for innovation and an effective contribution to the problem-solving and planning processes, but by adding dimensions to the FAST model we expand its usefulness in both the lead up to innovation and as a work-in-progress assessment tool during implementation. That is, the same FAST model can later be employed as an effective planning instrument that maps how a system is being improved progressively, month by month.

As discussed in previous chapters, FAST models are nondimensional. That is, when constructing the FAST model there are no sequence logics as suggested by time chronological order, such as "first we do x," "second we do y," "before we can do x, we need to do y," or "after we do x, then we do y." These are references to system or process terms. All we can ask of a function is, "Is it performed well?" The way in which we evaluate how well a function is performed comes from the measurements of solutions and the value we strive to achieve. This value is itself defined in a value study's *pre-event stage*, in which the strategic issues, problems, and opportunities are examined.[1]

PRE-EVENT STAGE

The pre-event stage is a project framing process where the conditions of a project's high-level business and technical objectives are discussed and issues decided that affect the outcome of the project. The decisions made during this process have a very direct effect on the theme, focus, abstraction level, dimensions, and basic function(s) of the FAST model. In fact, everything about FAST is decided here, including the makeup of the team that will build the FAST model and implement the approved corrective actions.

Among the issues discussed and resolved during the pre-event stage are:

1. *What is the justification and timing of the project?* The need to justify the project and the expected outcomes is necessary to set the direction and define the targeted success of the project. It determines if an "as is" or "should be" FAST model is required. The response will direct thinking toward developing a procedural (soft) or a technical (hard) FAST model, or both.

2. *What constraints must be overcome to achieve success?* This will identify real and perceived project constraints. It will also indicate the range and scope of the study area and the model's level of abstraction.

3. *What are the (five to nine) major attributes and their relative importance?* Addressing only the project goals may result in a conclusion reminiscent of "The operation was successful, but the patient died." The impact on major attributes must be considered in seeking innovative solutions, or the reason that justified the project's investment may be lost in the search for a narrow-based solution. Identifying the prime attributes and their relative importance to customer and producer in the FAST model will focus the team's attention on considering those high-valued attributes and critical project issues. One way to keep the high-priority attributes in focus is to dimension the FAST model functions by the project-selected attributes.

We explore these key issues and show how they can be articulated in this chapter.

A function is thus assessed through consideration of its performance and how that affects value at a strategic level. If a project were late, overbudget, and lacking in quality, when we ask "Is the function performed well by the current solution?" we would form a response with respect to attributes such as schedule, cost, and quality control, as a minimum. However, such dimensions are not contemplated until the FAST model is seen as being a valid account of that which it purports to represent. Until the validation of the FAST model is complete, we are uninterested in dimensions such as time, cost, responsibility, events, milestones, gates, calendar dates, and other units of measure, or metrics. Before we have a trusted FAST model we are only interested in the three logic questions *how, why*, and *when*, which help to shape the FAST model as an elaborate explanation of how something works (see Chapter 4).

It is only *after* the FAST model is complete and validated that dimensions appropriate to the issues described in the pre-event can be added to the model. These dimensions demonstrate how the performance of solutions contributes, either directly or indirectly, to strategic improvement. To reinforce this important concept, functions do not have dimensions. It is only when a method to implement a function is conceived that the implementation effects of the method selected mapped to that function can be dimensioned. We measure applied solutions and not functions. Therefore, the cost–function metric is the cost to implement a function in a unique way.

A properly dimensioned FAST model is a powerful technical and management analysis aid. Selecting the right metrics for dimensioning can be found by reviewing the information developed during the pre-event phase of the value study, where issues, expectations, and topic boundaries have been clearly defined. Selecting which dimensions to use depends on the definition of success as established by the project's stakeholders and customers in the pre-event stage.

FAST DIMENSIONING THEMES

There are many ways to display graphically the dimensions and metrics of the relationships between a real-world solution and an abstract function in a FAST model. Which approach or combinations of approaches to use depends on a number of considerations. The time available for dimensioning activities is often compressed without consideration of what such a reduction means in terms of higher levels of uncertainty and missed insights. The value of dimensioning the FAST model is that it links how things function to the results that management desires. Value is therefore a measure of the success of our innovations.

Many of the dimensioning techniques are equally effective with business systems and *soft study issues* (e.g., how to improve employee morale). They are just as effective with *hard study issues* as found in process, product, and equipment studies. What is more, a FAST model can represent both types of functioning in the same model. In this chapter we discuss some of the more popular dimensioning techniques and present dimensioned FAST models and business cases to illustrate the techniques and their effectiveness.

BUSINESS PROCESS AND SOFT ISSUES

In business process studies, time is often more important than cost reduction. Time to market, process time, and the time value of money are some key business issues where time *is* money. In addition to cost and time, FAST model dimensions include, but are not limited to, responsibility, accountability, budgets, personnel loading, expense allocation, determining value-added and non-value-added functions, process phasing, identifying stage boundaries, funding stages, paperwork flow, capital equipment assessment, allocating target costs, properly placing inspection and test points, paperwork reduction studies, establishing decision gates, identifying key milestones, positioning design reviews, and many others. In business process reengineering projects the FAST model can be used to determine direct responsibilities and show which personnel will be affected by the decisions of other groups. Goals can also be assigned to key functions.

Because functions are logically linked in the form of a dependency, the FAST model can display which department's goals are independent of other information sources. For example, the finish goods department does not need to receive information about the planned remodeling of the buying department. The organizational functions would not be on the same *how–why* logic path because they are about different intentions.

Similarly, FAST models can be used to illustrate which departments depend on the quality of information disseminated by other sources and which information is vital for a department's ability to achieve its own goal commitments. For example, the manufacturing department depends on the marketing department's sales forecast. They need to know, in advance, how many units to produce to plan effectively for production, so the functional relationships are either in the same means–ends path of *how–why* that explains core intentionality or they are in a strong causal relationship demonstrated by the use of the *if–then* or *when* logical operands.

SENSITIVITY MATRIX

As a method of dimensioning the FAST model, the sensitivity matrix method is favored where specific data are required for selected functions and activities (see Figure 3.4 as an example). The sensitivity matrix involves constructing a grid at the bottom of the FAST model or as a separate attachment to the model (see Figure 5.2). A sensitivity matrix simply provides a table in which we can easily see the cross-references between individual functions and those elements supporting the functions. Across the top of the grid are numbers that correspond to the uniquely numbered boxes of each function. The column down the left-hand side of the grid identifies the items being dimensioned, such as departments, disciplines, capital equipment, instruments, and process elements. The dimension symbols are located at the $X-Y$ intersections, where the item dimensioned relates directly to the numbered function. This tabular representation allows us to see which functions are sensitive to a proposed solution that is being considered and will now be highlighted in a case study.

FACILITY MANAGEMENT CASE STUDY

A medium-sized design and manufacturing company has experienced exceptional growth and the trend is predicted to continue. The need to expand their marketing and sales efforts while managing its growth led to the decision to outsource their manufactured components and focus on assembly and testing, supported by an aggressive marketing and sales strategy. The director of facilities was given the assignment to oversee a project to permanently house and equip the expected tripling of the market and sales staff within six months. To start the project, the director formed a core team consisting of those disciplines directly affected by this assignment, and with the team's assistance created a FAST model describing the assignment as functions. The FAST model shown in Figure 5.1 is a simplified version of the case study we just described. Note how clear our understanding becomes after reading through the FAST model. See Chapter 4 for details on how to read a FAST model. Although it has no dimensions yet, each function has been given a unique number to aid cross-referencing.

The primary issue common to administrative process value studies and their FAST models is determining who is responsible for implementation of the functions selected and who is moved to some kind of action as a result of the implementing actions of that function or group of functions. Developing administrative process FAST models is an example involving the assignment of responsibilities to functions and allows a means to innovate organizational designs through dimensioning and evaluation. Such a model enables management to check whether a more elegant organizational structure can be realized.

Determining Responsibility, Move to Action

After completion and validation of the FAST model (see Figure 5.1), the facility team constructed a sensitivity matrix displaying the affected departments in the left column of the matrix and corresponding function numbers across the top of the matrix (see Figure 5.2). We need to remember that the reason these items appear in the left-hand column is a corollary to the purpose of the FAST model and its links to strategic value.[2] In the FAST model shown in Figure 5.2, each function is assigned a unique number. The way the blocks are numbered is unimportant as long as they correspond to the function numbers on the sensitivity matrix at the bottom of the FAST model.

The departments affected by the project are displayed in the left column of the matrix. The core teams responsible for a successful project are shown in bold type.

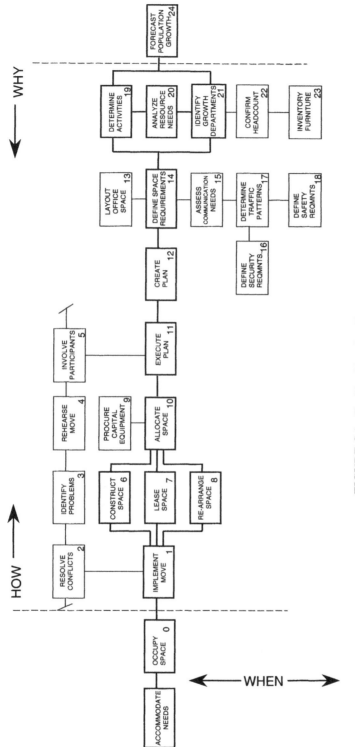

FIGURE 5.1 Facility Planning FAST Model

97

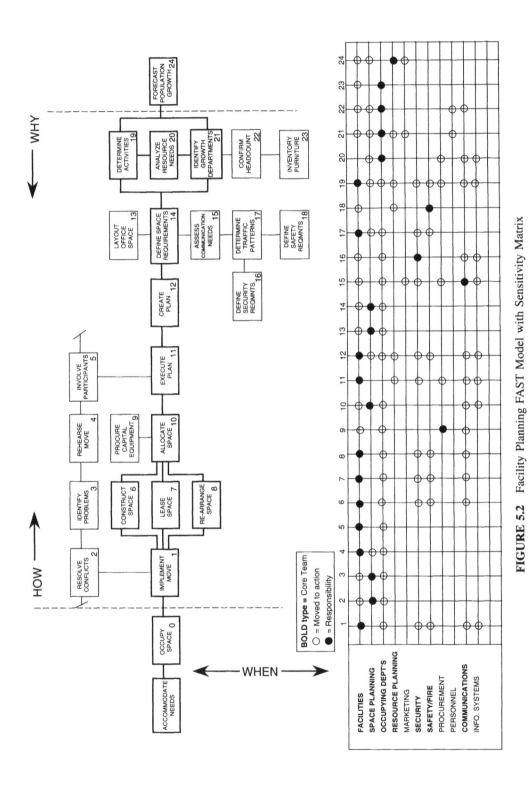

FIGURE 5.2 Facility Planning FAST Model with Sensitivity Matrix

The other departments are those supporting the facility team. The team then discussed who should be responsible for the individual functions and activities. A black dot at the matrix intersection signified the team's agreement on which department held responsibility to check that a particular function was performed satisfactorily. It was now that department head's responsibility to determine how best to implement that assigned function. At this point, readers should recall that the FAST model describes functions that need to be performed by solutions, but not what those solutions are. The facility team then determined which departments would be moved to action, or in support of the department responsible for implementation. A hollow circle at the department-to-function grid intersection indicated the selection.

INCORPORATING OTHER DIMENSIONS IN FAST MODELS

In addition to this example, other information can be added to the FAST model that improves its usefulness to the planning process (see Figure 5.3). These include posting key completion dates, locating the funding approval gate, and an estimate of the length of time needed to perform all functions between key functions. This can be illustrated by considering what functions need to be performed between creating the plan (see functions 11 and 12 in Figure 5.3) and actually occupying the facility (see functions 0 and 1) as well as achieving project completion after funding approval.[3] These and other dimensions added to the FAST model make it an effective planning tool.

Responsible department managers can clearly see which departments are affected by their decisions. As such, it provides a list for departments that should be consulted and informed as part of the planning and decision-making process. For complex systems such insights can help project planners to conceive better ways to build their schedules as a purposive logic rather than a task-oriented logic, and steers planning and subsequent activity. A comment often heard from responsible department managers is: "I knew I was responsible for that function, but I had only a vague idea of the departments affected by my decisions." By representing the system and how it works, FAST models aid practical ingenuity through building a shared and deep understanding of what needs to happen and why among people with managerial, scientific, and engineering knowledge.

RACI/RASI DIMENSIONING

In the facility case study above, we considered only two categories of managerial performance and in a digital way; that is, we assessed for each function whether someone either did or did not have responsibility or was moved to action (see Table 5.1). This series of decisions was coordinated in the sensitivity matrix (see Figure 5.3) so that we could consider functions individually and collectively in the same place on a FAST model. Now we augment the number of categories and build on the same approach to organizational design. Later we reflect analog and continuous scales and how the sensitivity matrix can help us to consider metrics such as time and cost.

RACI/RASI dimensioning is displayed in a sensitivity matrix form similar to the matrix shown in the facility planning model (see Figure 5.3) but offers more detailed information. RACI (R = responsibility, A = accountability, C = consult, I = inform)

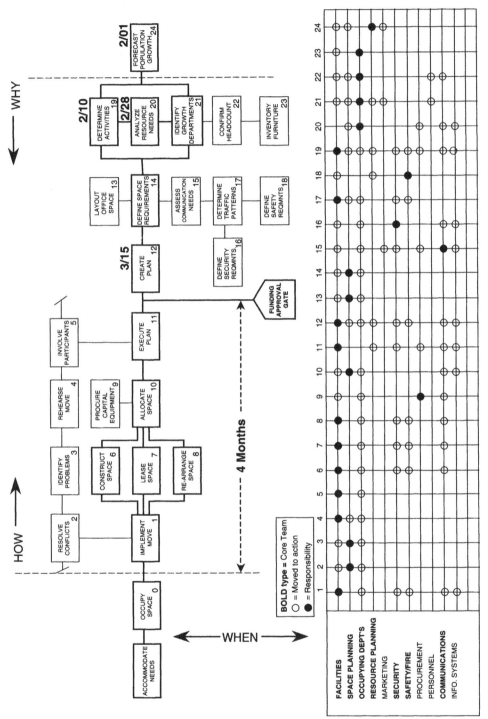

FIGURE 5.3 Facility Planning FAST Model as a Planning Tool

100

TABLE 5.1   Digital Choice Made in the Facility Planning Case Study

Function X	Yes	No
Moved to action		
Responsibility		

and   RASI   [R = responsibility, A = approval authority, S = supporting services, I = inform (or copied)] are two similar forms of determining a staff's association to selected functions with slightly different interpretations directed toward the same objective. In place of the black dot and circle (see Figure 5.3), RACI/RASI breaks the participation of staff assignment to a finer degree, and the particular distinctions are shown in the form of letters (i.e., R or A or C or I). The same would apply to RASI if that were the team's choice. At the point of intersection, a designated letter is placed indicating the level of participation in implementing the function selected.

We explain the background to a case study in the next section before actually looking at an example.

FAST AND ORGANIZATIONAL EFFECTIVENESS

Many companies recognize the need to reorganize the way they conduct business. In this age of accelerating communications technology, massive amounts of information can be quickly gathered, sorted, and used in making tactical and strategic business decisions. This readily available technology reduces the many management layers previously needed to segment and manage a variety of business-related information. The desire for lean operations and a higher production/staff ratio requires a hard look at the way that many companies function. The RACI/RASI dimensioned FAST model is an effective way to display, analyze, and reorganize the company structure to add value by improving the efficiency and effectiveness as well as reducing the cost of doing business. In doing so, the new organization will be more responsive to the market demands served by the company.

ORGANIZATIONAL EFFECTIVENESS CASE STUDY

Figure 5.4 describes the planning of a management team tasked to reorganize a software system architecture department in order to improve its responsiveness to internal and external customers employing their services. The parent company develops high-tech equipment for the government. The department has the capability to fully develop software in support of the products produced. However, this software support service was not fully captive. Product development program managers had the option to subcontract their software needs if the internal software department could not fully comply with the program manager's software requirements, development schedule, or cost expectations.

The head of the software development department was concerned that more and more product development program managers were outsourcing their software development. A lack of responsiveness to the program manager's needs appeared to be the

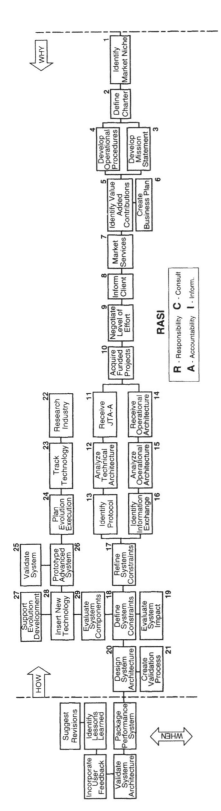

FIGURE 5.4 Software Architecture Team RACI

principal reason for outsourcing. The head of the software development department decided to reorganize his department to respond to the shortfall. The FAST model shown in Figure 5.4 is one of four development management systems operating within the department. All four management systems were studied, but here we focus on principles related to the dimensioning of just one management system, the software architecture team.

Model the Future or the Present?

Following the pre-event discussions, the team facilitator opted to lead the team in constructing a "should be" (i.e., the future) FAST model rather than display the current organization structure (i.e., the present). The "should be" approach to building a FAST model is best used when the client encourages a total restructure or "clean sheet" approach (see "'As Is' Versus 'Should Be' Models" in Chapter 4). When selecting the "should be" approach, it is best to include *wild cards* on the team. A wild card is a highly technically qualified team member who doesn't have a vested interest in the outcome of the study. The different perspective helps the organization's staff, also serving as members of the study team, to break through their paradigms and explore unconventional but valid approaches they may not have considered.

Following the completion of the FAST model, the management team agreed on the following definition of RACI:

- R: responsibility, the single approval authority for the function's task
- A: accountability, those making up the function's implementation task teams
- C: consult, those coordinating and supporting the function task teams
- I: inform, those required to act as a result of the function task team's output

Once again we remind you that value is ultimately recognized as an outcome. The accuracy of the definitions assigned and agreed in a RACI matrix relates to how well the chains of command and communication channels are designed. This ability to assess management capability will aid the software architecture team's department to achieve its ambitions to become a better provider of solutions to the internal market. One has only to look at the number of functions with more than one "A" assigned to realize that this team has confused accountability issues (see Figure 5.4). Here we would inquire how tough choices are made with respect to the solutions used to perform those functions with more than one person in leadership roles. We would also look at how those with responsibility deal with overlapping tasks and hand over issues. The FAST model with a RACI allows us to inquire far deeper than an organization chart would allow.

As described in the facility planning study (see Figure 5.2), in the sensitivity matrix the left column of the grid identifies the department's organizational elements affected by the way things are performed Across the grid's top are the discrete numbers corresponding to the numbered functions in the FAST model. At the points of the grid's intersection, the management team determined the function's contribution to the project by selecting the appropriate letter in the RACI. Once this was achieved, the existing organizational chart was examined to see where lines of communication and roles and responsibilities might require redesign. We discuss this further in the next section.

Incorporating Additional Dimensions

Although RACI is a powerful way to relate staff assignments to functions, other dimensioning schemes can add valuable information to the project FAST model.

Performance Color Coding   Following the resolution of the RACI selection, the team compared the performance of their "should be" FAST model to their current performance. Using a color code (see Figure 5.5), the team used color to describe the current performance level of the functions. Figure 5.5 uses symbols in place of color because the figures are in black and white. The actual FAST model used red to show those functions not currently being addresses, yellow for those functions performing poorly, and green for those functions performing well. The color-coded FAST model gave a clear indication of what topics should dominate the next stage, the creative phase of the innovation process within the value study.[4]

Clustering   Functions were grouped to identify specific organizational units responsible for the performance of those functions. In the process of evaluating the current organization's performance against the "should be" model, the management team identified suborganizational elements by clustering related functions and identifying their role in the department's organizational structure. In doing so, it allowed each department's inputs, based around management systems, to form subteams and address the issues of concern in greater detail.

Selective Lower Level of Abstraction   When the management team identified the functions relating to the technology insertion cluster, they determined that more detailed information describing function 30, *Prototype Advanced System*, was necessary for analysis. The inverted "V" with the letter "A" symbolized bringing that function to a lower level of abstraction and expanding its functions. *Prototype Advanced Systems* was examined on a lower abstraction level, because of its interest to the team, and displayed in the upper right corner of Figure 5.5. Only by lowering the level of abstraction of selected functions, rather than moving the entire model to a lower and more detailed level, can the team focus its attention on key issues instead of wading through an overly complex FAST model.[5]

PRODUCT- AND EQUIPMENT-BASED FAST MODELS (ARTIFACTS)

FAST models in this category are involved with hardware, such as product improvement, product development, capital equipment, and construction projects.[6] The same dimensioning techniques and FAST modeling principles apply equally well when constructing an artifact type of FAST model. The difference is that in "as is" artifact FAST models, a physical product (i.e., solution) already exists and can be examined, dissected, evaluated, and analyzed in a more objective manner. In the facility planning case study examined earlier in this chapter, the alternative was shown where intangible soft issues as is common in management systems enacted by people was modeled and so may be tainted by subjective interpretations and personal agenda.[7]

Sensitivity Matrix in Product (Artifact) Analysis

The sensitivity grid constructed for FAST models involving equipment and hardware products is similar to the organization effectiveness studies described above. The staple

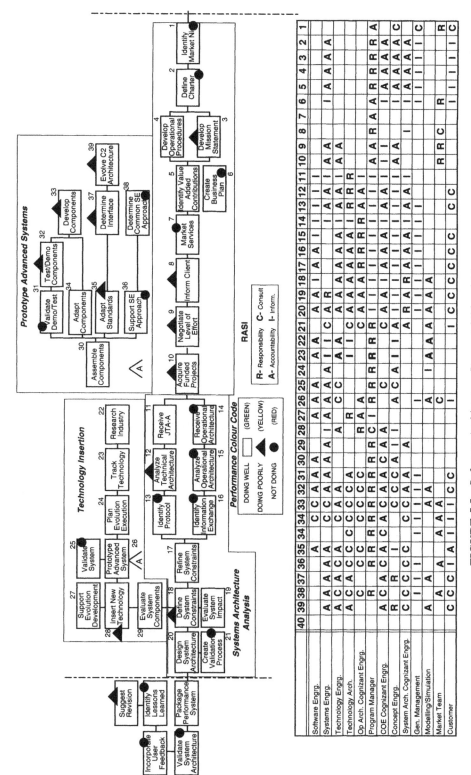

FIGURE 5.5 Software Architecture Team: Additional Dimensions

remover example described next illustrates the use of the sensitivity matrix for artifact-type studies. This simple example is selected to direct the focus on the process rather than on the example's part costs.

STAPLE REMOVER CASE STUDY USING FAST WITH THE SENSITIVITY MATRIX

The sensitivity matrix at the bottom of the staple remover FAST model (see Figure 5.6) shows the contributions of the product's parts (left column) to the functions supported by those parts. By allocating the cost of the parts to their related functions, a cost per function relationship emerges. Where multiple parts serve a function, the total cost per function identifies which *functions* to address in searching for ways to improve the value of the function–cost relationship. Dividing the part cost by the number of functions it serves is a blunt way to allocate part cost to functions, which may suffice in time-constrained episodes. A better approach is to estimate the percent part cost to those functions it serves. Even so, the allocation is often very subjective because a solution performs several functions, and to untangle them in a way that allows objective cost allocation is a judgment call.[8]

The intent of this dimensioning method is to establish a comparative relationship of the cost to perform functions rather than creating an accurate cost estimate. Therefore, depending on project complexity, either a simple amortization or an estimate of the percent component cost to perform that function is useful for the purpose of comparing relative costs per functions. In the staple remover example, the left column in the grid identifies the components of the design to fulfill the basic function *Remove Staples*. The next column displays the total installed cost (TIC), in cents, of the parts. The numbers across the top of the grid correspond to the individually numbered functions shown in the FAST model.

Determining Component Function–Cost Details

With existing products, improvement begins with an "as is" study. A FAST model is constructed that displays the function contributions of each major component. This is necessary for product improvement because it reveals the accumulated cost contributions of the components selected to perform those functions. These solutions are then questioned and analyzed by the VE team's effort to increase value. Questions such as "How else can we *Transmit Force?*" open discussions that lead to innovation. To establish the function–cost relation of parts, each physical part is analyzed to determine which functions that part serves or supports. The cost of the individual parts is then amortized by the number of functions served, and that value is placed in the grid's part–function intersection. Determining the function–cost value can be achieved by estimating the percentage of the part cost that supports a function, as in Figure 5.6, or as often used in more complex studies, a straight amortization procedure is used.[9]

The object of the process is to determine the appropriate cost per function, *not* to create a detailed cost estimate. Examining the appropriate cost per function and the number of parts required to support a function identifies where the team can direct its creative efforts to reduce cost and improve value. Functions 7, *Position Wedge End*; 15, *Hold Fingers*; 17, *Squeeze Fingers*; and 18, *Resist Spring Force*, have no cost per function, because although they appear to be functions, they are procedural steps, or

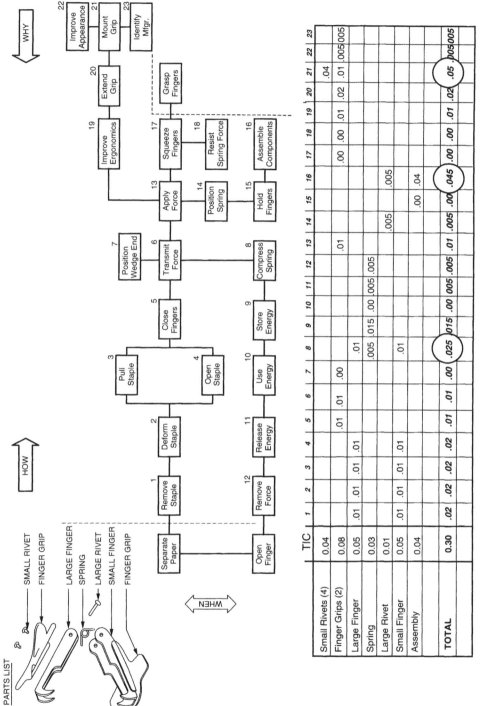

FIGURE 5.6 Staple Remover Example

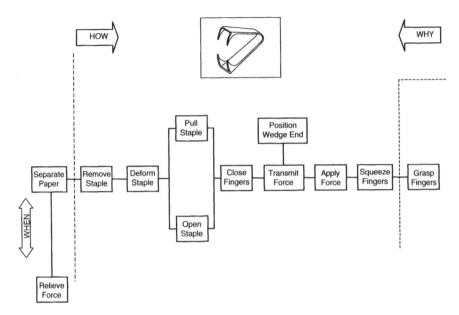

FIGURE 5.7 Proposed Staple Remover

as in function 18, outcomes that are not served directly by any single component. (In this instance, the user's hand performs the function.)

The circled cost per function totals were identified as functions to be addressed in the creative phase because of their relatively high cost and because more than one part supports the function selected. This realization indicates that a possibility may exist to integrate the performance of functions in a new solution.

Proposed Solution

Figure 5.7 is a solution offered by a team during a FAST modeling training program. The proposed configuration is a single part that was stamped and formed into the staple remover configuration. All the individual parts and their associated costs have been eliminated. After forming, the part was heat-treated to incorporate spring steel properties, which performed the functions of the spring in the previous design. The part was then plated to improved aesthetic and perceived value. The cost of the proposed design was about 25 percent of the original design. As the purpose of value studies is to add or improve value, the key question that must be resolved in evaluating the proposed design is, "Can you sell it?" If the proposed design lacks customer appeal or perceived value, it cannot be sold regardless of the reduced cost and price; value is recognized in the strategic context! The next issue to be resolved is: What innovative features and functions must we add to the proposed product to increase perceived value, at a competitive cost?[10]

PIPELINE CASE STUDY USING THE SENSITIVITY MATRIX

A case study involving the construction of a gasoline pipeline project is described in FAST model form (see Figure 5.8). In this episode, revenue was determined in a license agreement that placed a limit on production volume, so the strategic challenge was to leverage profitability. The VE task team that developed the FAST model was charged with reducing the progressively increasing budget for the project. At the time the program manager decided to conduct a value engineering study. The project was completing the select phase of a value creation program,[11] where the path of the pipeline, size of the pipe, number and locations of booster stations, valves, and a general approach was selected schematically and costs estimated. The intent of this study and of the FAST model was to identify the major function costs. Those key functions were the topic of a more detailed study conducted on a lower level of abstraction, to reduce the cost to achieve the functions while maintaining the project's output expectations (e.g., x gallons of gasoline distributed per day).

The sensitivity matrix is similar in form to the previous examples, where the number of functions was supported by the amortized major project expenses. The left column in the sensitivity matrix displays the project's major budgeted expense categories. The second column displays the budgeted total cost of those major expense categories, in millions of dollars. Using the simple amortization process, each category's budget is apportioned to the function supported by that budget item. The "total" row sums the relative cost to perform the functions identified in the FAST model. This is an example of a high-abstraction-level FAST model. To keep the team's focus on the project's major issues, the team opted to number and identify only the functions on the major logic path, the rationale being that since the functions shown in the *when* direction are a direct effect of the major logic path functions, the budget allocation will account for those functions not numbered.[12]

OTHER CASE-SPECIFIC DIMENSIONS

FAST models are not limited to the dimension examples discussed. Other dimensions noted in the pipeline study (see Figure 5.8) are functions selected for study in the creativity phase, stages' gate boundaries (e.g., project phases such as Define, Execute, and Operate), key stage dates, and the location and purpose of the major approval gates. A project manager would be well advised to learn who the gatekeepers are, and the requirements for passing through the gates, well in advance of the event. A well-prepared project manager entering a gate review will avoid needless delays that could disrupt the workflow and increase unplanned expenses of the project management process. Supplementing the sensitivity matrix with other case-specific dimensions translates the FAST model into a more useful tool and broadens the scope of value-improving opportunities. Because functions are an abstract representation of reality, a dimensioned FAST model also becomes an effective planning map.

BUDGETING OPERATING EXPENSES AND THE SENSITIVITY MATRIX

Since department budgets are based on forecasted activities, and activities are the way that functions are implemented, an interesting application of the sensitivity matrix dimension process is to evaluate a department's budget against its functions rather than its activities. Starting with the charter and mission statement of a department, create a FAST model displaying the functions of that department. When complete, create a sensitivity matrix. In the left column, list the projects and operating activities

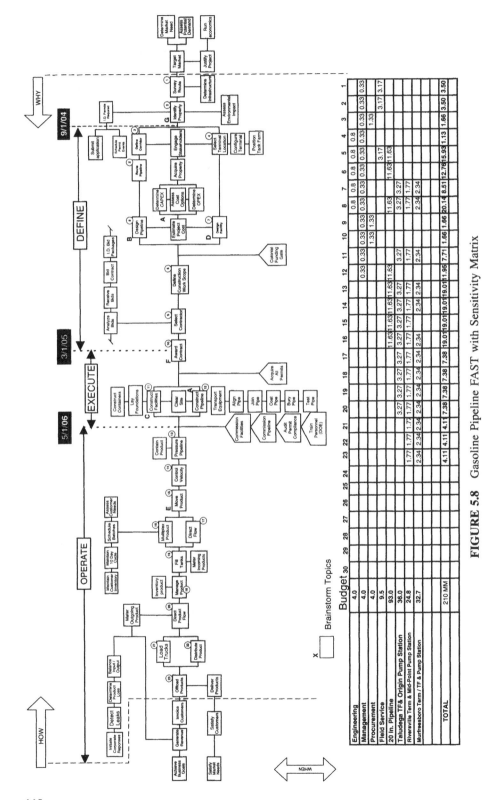

FIGURE 5.8 Gasoline Pipeline FAST with Sensitivity Matrix

110

that justify your budget. A second column should show the budgets assigned to the projects and departmental operations. The next step is to amortize the budget against those functions to be performed by the solutions we are examining. Examining the total budgeted expense to perform related functions will highlight where attention and analysis should be focused for trimming budgets and activities that do not support the department's charter. The analysis will also identify which planned activities are not within the charter of the department but in support of other department's tasks or objectives.

Using the FAST model and sensitivity matrix technique to examine budgeted expenses will result in a leaner department while maintaining, or improving the charter functions of the department, which justify that department's existence. By reducing activities outside the department's charter and reinvesting some of the savings to strengthen the department's major functions, the department will become more productive with the same resource base, and thus improve.

CLUSTERING FUNCTIONS

This dimensioning technique involves drawing boundaries around groups of functions to represent subsystems. Clustering functions is a good way to illustrate value improvement targets and to assign *design-to-cost* (DTC) targets for new design concepts.[13] For value improvement, a team made up of marketing, design engineering, cost accounting, and production would develop an "as is" product FAST model, cluster the functions into subsystems, allocate product cost to the clustered functions, and assign target costs and at the same time consider how to improve income. During the act of creating the model and the conversations that flit across disciplinary knowledge boundaries, functions that customers view as being important can be identified. Identifying the functions that customers value more than other functions opens the way to opportunities for significant value improvements in design and production. The use of dimensioning functions for design-to-cost and concurrent engineering (CE) projects can have dramatic results. The process of developing and dimensioning the model by an interdisciplinary FAST team will raise the level of understanding of the team members and bring them to a consensus as to where innovation is needed and achievable.

EXAMPLE USING CLUSTERING

A cost reduction example of a clustered function model is illustrated in Figure 5.9. The example displayed in Figure 5.9 is a design-to-cost value engineering assignment to design a new air-cooled generator where cost is treated as an engineering requirement.[14] Using the "as is" product as the base, a FAST model is constructed showing the major functions of the product. Selected functions are then clustered and identified as components or assemblies of interest in achieving the design-to-cost objective. The dashed lines on the model border the various subsystem clusters that comprise the product. Each cluster represents a study topic. Following the completion of the model, the subsystem clusters were divided among three subteams. The team member's objective was to achieve or beat the assigned target costs. Where necessary, the teams expanded the clusters by moving that segment of the model to a lower level of abstraction. This revealed the subsystem clusters in greater functional detail. The VE team then

112

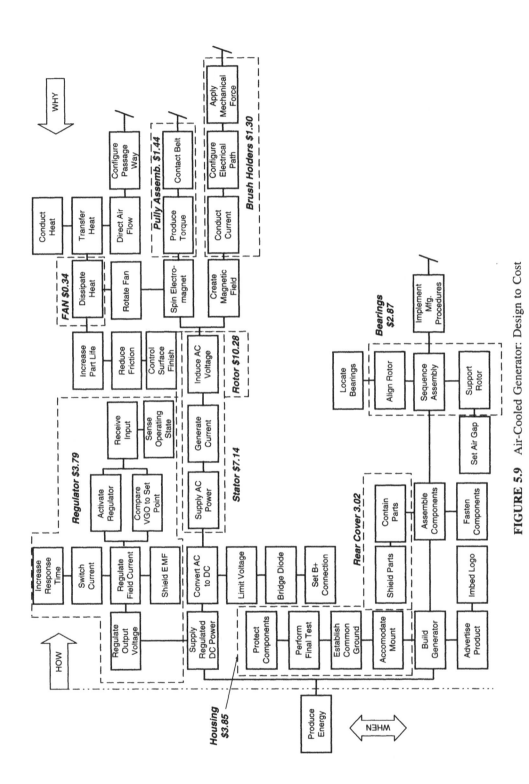

FIGURE 5.9 Air-Cooled Generator: Design to Cost

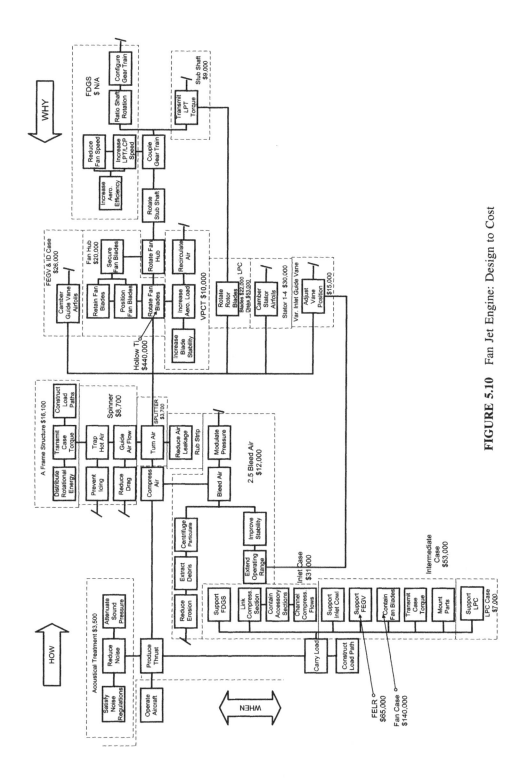

FIGURE 5.10 Fan Jet Engine: Design to Cost

addressed the clustered functions and studied how better to achieve the design-to-cost goals by addressing the major functions within each cluster, not how to produce the part or assembly for a lower cost.

During the creative phase, the VE team may move a cluster to a lower level of abstraction and create an expanded sub-FAST model to study the part or assembly in more detail. Also studied are ways to eliminate parts or assemblies by combining clustered functions within a new single component in an innovative design. It is important to note that the part and assembly costs are targets. As such, the proposed design may show some clusters exceeding, and others lower than, the target cost. This is acceptable provided that the total product design-to-cost target cost is achieved or the cost is lower than the target.[15]

Product size and complexity should not be a deterrent in using this effective FAST modeling technique. Figure 5.10 is an example of a design-to-cost value engineering study to develop an improved and lower cost for a fan jet engine. As in the air-cooled generator example (Figure 5.9), the FAST model that describes this low-pressure compressor (LCP) fan jet engine used the clustering functions approach to establish design target costs. The clusters were assigned to four VE task teams, which then sought to lower the level of abstraction and expand the number of functions considered for a more revealing view of what actually happens. Each team focused on achieving the targets for their assigned clustered subsystem. They were encouraged to reach beyond best-in-class goals. However, they were also instructed that a team must not meet or exceed their assigned goals at the expense of another team. This forced the subteams to network and address the bigger total product target cost goal.

CLOSING REMARKS

In this chapter we focused on dimensioning the FAST model to make it a more dynamic analysis instrument. Major attention was devoted to the sensitivity matrix as a way to assign key measures or dimensions to individual functions. This technique works as well for business systems and soft issue studies as it does for hard process, product, and equipment studies. Also discussed was the clustering dimensioning technique. Clustering is simpler than a sensitivity matrix to construct, but properly employed, the techniques can be equally effective.

There are many variations to the dimensioning techniques presented, and some additional schemes. To cover all the possibilities would far exceed the limits of this chapter. The reader is therefore encouraged to experiment with the techniques discussed and with new approaches to dimensioning the FAST model. As more practitioners use, experiment, and develop new dimensioning techniques, the FAST model will expand in usefulness and diverse applications. The key lesson to remember is that because a product's value is realized at the strategic level, the FAST model must form a conceptual link between how something works and how that's so useful that a customer will pay to enjoy the value provided. The other key lesson is that although we break things down into parts, it is their synthesis that yields a fully functioning product, service, project, artifact, and so on.

In the next chapter we explore the link between the strategic context and the operational context. This will allow us to align customers' value with what the organization is capable of delivering.

CHAPTER 6

ATTRIBUTES AND THE FAST MODEL

In the last few chapters we have focused on how a FAST model is constructed from a technical perspective. We need to widen this into a broader outlook and one from which organizations can achieve rapid innovation systematically. In this chapter we explore how a FAST model is aligned with complex problems involving tangled issues, mixed objectives, and the need to find a way to make sense of things before decisions are made and actions taken. The discussion in this chapter prepares the way for subsequent chapters in which we explore the strategic value of FAST modeling and how it can help to design and implement strategic competitive advantages. Let us begin by understanding the relationship between context and priority and start by explaining how the pre-event phase directs the direction of the FAST model and those who are selected to develop it.

Customers are faced with choices. They don't always have the time to inquire into functionality, and in many cases may lack the knowledge to do so meaningfully. For example, when someone buys a television set, he or she doesn't spend time discussing the circuit designs but simply wants to know whether it will perform in such a way that the person will derive a sense of satisfaction in having spent money to purchase and own the set. So there is a relationship between performance and functionality. *Performance* is how customers assess the efficiency and effectiveness of the engineering solutions selected to perform functions. *Functions* are about the essential contributions needed to make a thing valuable. We measure improved performance against the attribute's past performance.

Given that rival products or services perform differently, we as innovators need the ability to compare them. We do this with a concept called an *attribute*. Identifying the major attributes that affect the performance and perceived customer value of a project under study is as, if not more, important than establishing meaningful project goals. The

Stimulating Innovation in Products and Services: With Function Analysis and Mapping, by J. Jerry Kaufman and Roy Woodhead
Copyright © 2006 John Wiley & Sons, Inc.

solutions selected to perform functions cause the effects we measure "on" attributes. This is why attributes and FAST models are inextricably linked. Determining a project's performance attributes occurs during the pre-event process, discussed in Chapter 5.

An attribute is a range of inherent characteristics of a product or process that affects its value perception in the eyes of the customer. This allows us to compare the effects of various solutions used to perform functions on the FAST model. That value perception can originate from the producer's internal customers, the external market customers, or both. Figure 6.1 shows an example of time as an attribute, where t_0 is "now." Imagine that we are considering different solutions to the function *Deliver Product*. How could we identify one idea as offering more value than another without some kind of yardstick? It is important to see that an attribute is a range of possible specific points. In the case of time, the attribute is the line on which continuous data can be represented. When we look at a particular or specific situation, we think of a single point on such an attribute: for example "two weeks." Such a specific point is only meaningful if other points, such as "one week" or "last week," actually exist within this framework. So an attribute has characteristics, which are particular points along the attribute (see Figure 6.1). As such, the concept "two weeks" is a particular characteristic that is simply one point on an attribute. When considering subjective assessments of value such as quality, we could discuss noncontinuous data in the same way by differentiating one characteristic from another in some progressive way from "low" to "high." Let's say that we are hired to improve the delivery performance of a company selling books on the internet, and that this currently takes about one week. This particular characteristic of the service is a starting point for our thinking, so the notion of one week becomes part of the base case. If after benchmarking we see "best in class" rivals offering shorter delivery times and as a consequence we want to achieve a reduction to three days, we target another characteristic that is an improvement on the base case. When we target a particular characteristic, such as delivery in three days, we are setting a goal.

The problem we face in Figure 6.1 is that we need a closer look at attributes rather than having to consider impractical extremes, so we "bound" the attribute by placing scoping lines on it. Rather than considering time from minus to plus infinity, we chose a more meaningful focus, such as a three-day period, and perhaps start talking in terms of days of improvement. Another problem with Figure 6.1 is that we are unsure how to compare different types of attributes, such as "time" and "quality." We will now elaborate on our approach so that such problems can be overcome. The major distinguishing feature of an attribute is that it is acceptable within a range of goodness

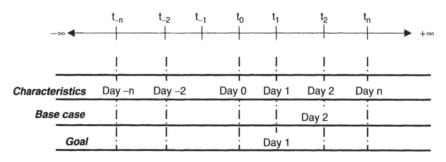

FIGURE 6.1 Time as an Attribute

(see, e.g., Figures 6.2 and 6.19). We bound attributes with respect to the necessities of decision making. At one end is the "least acceptable" limit that customers would entertain. At the other end is a reflection of "excellence" that a team should strive to surpass.

We often show the attributes on a *range of goodness table* (e.g., Figure 6.2) but can sometimes show them on the actual FAST model as the dimensions in a sensitivity matrix (see Chapter 5). In a project that sought to innovate internet routers, where milliseconds of time is key, we even placed individual attributes under each function and mapped how different solutions, hardware, and software performed at both the total product level and at various component levels. A similar example, which shows both a range of goodness and the attributes on a FAST model, is shown in Figures 6.2 and 6.4.

We often select a collection of rival products or services against which to set benchmarks. These products may well perform the same functions and have similar FAST models but will use different solutions, or ways, to perform those functions. For each attribute we identify what the best in class achieves and then determine what would be better. This may be difficult to achieve but is feasible. If we have, say, four rival products, product 1 may be best in class for attribute A, product 2 for attribute B, and so on. This means that we determine what best in class could be and then position rival products so that we know how best to compete against best in class. Let us look at a practical example. An ideal project completion goal for next year is established as May 15. Selecting project completion as an attribute would display three points, the target completion date (i.e., May 15), the low end, or the worst acceptable date of (say) August 15 and a "stretch" goal. Let us assume that following a benchmarking project we ascertain that best in class could achieve this by April 15, one month sooner than our target, so we set a stretch goal that surpasses our best rival by setting an ambitious target of April 5; we always try to push the performance envelope. This five-month range, April to August, is the *range of goodness*. Any completion date that falls within that range is acceptable to the project management team. However, the managers would encourage the project team to try for an earlier target date without adversely affecting other critical requirements. If the project's contract offers substantial incentives for early completion and sizable penalties for exceeding the target completion date, that attribute would be very sensitive to the success of the project. It is important to see that the attributes are linked to a notion of customer value. If we improve one attribute, we may degrade another. They correspond collectively to a singular notion of project value.

Once functions are mapped on a FAST model, we can ask "How else can we perform this function?" and the attributes should steer the direction of our innovative efforts. Therefore, constructing a FAST model that describes the project is a very useful way to address the attributes. If time is a project concern, the FAST model can be coded to identify time-dependent functions and analyzed to find which functions in the major logic path are time constrained. Identifying the activities designed to implement those functions will lead to studying ways to reduce process time. That is, we search for different solutions that lead to schedule improvement and other sources of value. Examining the other functions that are logically dependent on the now-modified functions (read in the *how* direction), we will quickly recognize that many other functions and their current solutions will be affected by the proposed changes. This "thinking things through" will test that changes made to some functions don't create bigger problems somewhere else in the system; we must never lose sight of the systemic view of the completely functioning system.

Remember that a process does not always have to have a needed function and so may be removed with no detrimental effect on overall value. Recalling the difference between a FAST model and a process model, a FAST model identifies what has to be done. A process model describes a particular way to do it. Therefore, a process model is someone's concept, or solution, for the way to implement functions in the FAST model. To isolate the time-sensitive constraints, it is best to identify the functions that are affected and then examine the way that those functions are implemented; it is not the function that is timed but the solution selected to perform the function. By first examining the FAST model, the decision to modify, combine, or eliminate functions, or just change the way the function is implemented (i.e., a new process is substituted) is a way to correct the problem rather than to "fix" symptoms.

DEFINING AN ATTRIBUTE'S RANGE OF ACCEPTANCE

What we are doing is linking the FAST model, the solutions used to perform functions, and the measurement of the solutions on attributes to link FAST models to value. Suppose the project team said that they could come close to the stretch goal if some other requirements were relaxed, such as capital expense, the theory being that if we spend more money, we can improve schedule. The project manager's response may be that the other requirements must not be traded to improve the schedule. "In fact," the project manager may add, "I would like to see us deliver higher performance than specified." Based on the project manager's decision, the project team could not be faulted for believing that the other requirements are regarded as more valuable than time or the schedule. If the project team shifts to improving performance (e.g., reducing capital expense) as a result of the project manager's comment, rather than finding ways to improve the schedule, the opportunity to reap additional benefits may be lost. How many attributes should be addressed? What are the key attributes and their metrics? What is the relative importance of these attributes? These are questions that should be resolved before launching any project.[1]

A project has many attributes, but to manage them effectively the project leader should take the time with the project team to identify six to nine key attributes and articulate the characteristics along a range of goodness. By doing this as a collective endeavor, the project team becomes more aware of what success looks like in the eyes of senior management. Here we are using management as an internal customer. It is very important that the customer be represented in these discussions, as the customers will ultimately determine success. To assume to know customer preferences is dangerous. If the customer is actually a wider constituency of different stakeholders, more formal approaches to inquiry, such as the research methods participative inquiry and focus groups become necessary. Figure 6.2 is an example of displaying attributes and their range of acceptability or goodness.

The case study represented in Figure 6.2 is a proposed facility to produce polyethylene in pellet form. The pellets are the feedstock for other industries using polyethylene in their manufacturing process. The performance of a similar smaller-scale operating plant was used to select and predict the outcome of the attributes. The challenge was to scale the process and improve it.

Polyethylene Production ★ *Current Position*

ATTRIBUTES	1	2	3	4	5	6	7	8	9	10
Capex	**100**			**90★**						**70**
Opex	**100**	**97+★**								**70**
Schedule	**26**					**23★**				**21**
Constructability	**1**							**8★**		**10**
Operability/Availability	**95**						**98★**			**99**
Power Consumption (MW)	**32**				**28★**					**26**
Technology Certainty	**High Risk**				**Moderate Risk**				★	**Low Risk**

FIGURE 6.2 Attributes' Scale of Goodness

After long discussions and an initial problem framing session, the project management team identified six attributes they considered important to produce and distribute a product competitively. That is, the way they would recognize a better project would be assessed by measuring solutions selected to perform functions in the FAST model on these attributes. The attributes were:

- *Capex*: capital expense as a percentage of the current estimate (U.S. dollars)
- *Opex*: operating expenses as a percentage of the current estimate (U.S. dollars)
- *Schedule*: months from sanction to fully operational status
- *Constructability*: subjective; based on a benchmarked plant
- *Operability/availability*: based on usable capacity and throughput
- *Power consumption*: total units (megawatts)
- *Technology certainty*: (this attribute dropped out in the paired comparison analysis to determine attribute ranking; see Figure 6.3)

	B	C	D	E	F	G	H	I	J	Description	Score	Percent	
A	B1	A2	A2	A1	A1	A1				Capex	7	25	A
	B	B2	B2	B2	F1	B1				Opex	8	29	B
		C	C1	C1	C1	C1				Schedule	4	14	C
			D	E1	G1	D3				Constructability	3	11	D
				E	E2	E1				Operate/Availability	4	14	E
					F	F1				Power Consumption	2	7	F
						G				Technology Certainty	0	0	G
							H						H
								I					
										Total	28	100	

FACTOR

1 Small value advantage
2 Moderate value advantage
3 Significant value advantage

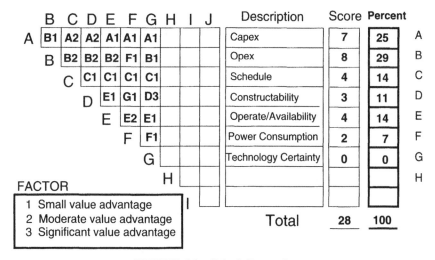

FIGURE 6.3 Paired Comparison

To determine the relative ranking of the attributes, each attribute was compared against the other attributes to determine the relative ranking. This enables us to represent a combination of attributes that reflect a singular all-around performance level to which we attach notions of value. Because we do this, we can consider trade-offs between the attributes. For example, if we used the cheapest material for the pipes that would carry various acids, the corrosive effects of the chemicals would mean that the pipes would have to be replaced in three to five years with a major facility shutdown. If we used an expensive material for pipes, such as zirconium, the pipes would not need replacement for at least seven to 10 years and a number of revenue-losing plant shutdowns would be avoided. So we could trade off the cost of material as capex against replacement cost as opex and lost revenue from plant shutdowns, provided that the traded attribute did not fall below the worst acceptable case.

RANKING ATTRIBUTES

Since all the attributes are important to the success of a project, attributes are ranked and positioned by determining which attribute requires more attention relative to the other attributes and their importance to the customer: in other words, given the state of the project today, which attributes, if improved, would lead to greater value. The ranking is therefore case specific and should not be used as a template for other projects. Similar projects may use the same attributes, but each project will have unique characteristics which will change the ranking of the attributes relative to that similar, but not the same, project status.

In the paired comparison exercise (see Figure 6.3), one attribute almost always score "zero." This does not mean that the attribute, in this case technology certainty, has no value. It simply means that of the seven attributes selected, an improvement in this attribute was not viewed as being able to yield significant value for the project as envisaged today. In our case study, technology certainty was dropped because its position on the goodness scale (9) indicated that the project would use tested process technology, representing low risk. Therefore, this attribute would not require much attention as long as it was not degraded substantially.

The selection score of each attribute is totaled and summed. In this case study the score is 28 (see Figure 6.3). Normalizing the numerical attribute score converts the score to a percentage. The result indicates that opex, capex, schedule, and operability/availability are the four high-value attributes of concern. By ranking the attributes, the decision of which two attributes to trade if a trade situation should arise would favor the higher-valued attributes, provided that the attributes traded do not fall below 1 on the attribute goodness score.

Trading attributes is not a zero-sum game. That is, attributes can be improved without adversely affecting another attribute, and this is a creativity ambition for the team. Remember that in this case study we have selected only seven attributes out of the many attributes available in the project, because these are the key value drivers.

INCORPORATING ATTRIBUTES INTO A FAST MODEL

Placing the attributes in a FAST dimension matrix (see our discussion of the sensitivity matrix in Chapter 5) will indicate which solutions attached to functions are attribute sensitive (see Figure 6.4). While the Figure is difficult to read because of the very small font size, we direct your attention to the way the FAST model is dimensioned, not the polyethylene process, and hope that the gist of what we are saying is clear. A FAST model can have multiple dimensions on different sensitivity matrices to analyze a variety of conditions that are solution-function sensitive. The value of multiple-dimensioned FAST models is that they can display seemingly different management issues on a single FAST model. In so doing, key functions can display their influence on different systems.

The FAST model displayed in Figure 6.4 can exhibit the list and cost of capital equipment, amortized to show the functions served by each piece of equipment (e.g., extrusion, with a total installed cost of 50.6 percent). A RASI matrix can be displayed and identify those responsible and accountable for each phase of the process, and as shown in Figure 6.4, another type of matrix can identify those functions that are attribute sensitive. The attribute matrix is shown to call attention to this topic, and we have already discussed the RACI matrix in Chapter 5.

The clustered functions in the FAST model identify the major elements of the process (see dashed lines around groups of functions that are clusters). Also shown is the percentage of total installed cost committed to the budgets for those process elements represented as clusters of functions. These percentages were left in the model as an example of different ways to show a multiple-dimensioned FAST model.

Knowing the project-sensitive attributes and their relative rank, the project team can select the functions that have a large effect on the attributes as focused areas for improvement, as they are value drivers. As noted earlier, by identifying the other functions to the right that depend on the targeted function, the collateral effects of proposed improvements can be assessed (chase through the *how–why* and *when* logic paths). Doing so would ensure that an improvement to one function does not adversely affect other functions or solutions that we want to retain. It's all about using the FAST model and attributes to cross-reference "What is to be done?" and "What will happen if we do it?" and so opens up a dialogue for innovation.

LINKING ISSUES OF CONCERN TO A FAST MODEL

In the following sections we show how the preliminary investigations steer the direction in which the FAST model will be developed. It is important to understand that underlying FAST modeling is a view of a functioning universe. The Moon affects the tides and the Earth circumvents the Sun because of functional relationships involving gravity, speed, distance, and so on. This can be expressed as a dynamic relationship such as:

- If function X, then function Y.
- Function Y is possible only because of function X.
- Therefore, function X has a functional relationship that is necessary to function Y.

People have created their own systems, such as the economy, which reflect a logical link to cause and effect. The key difference is that whereas nature functions mindlessly and arguably has no intentionality, humans seek to change things in nature to their own advantage (e.g., converting a mountain into stone with which to build houses).

122

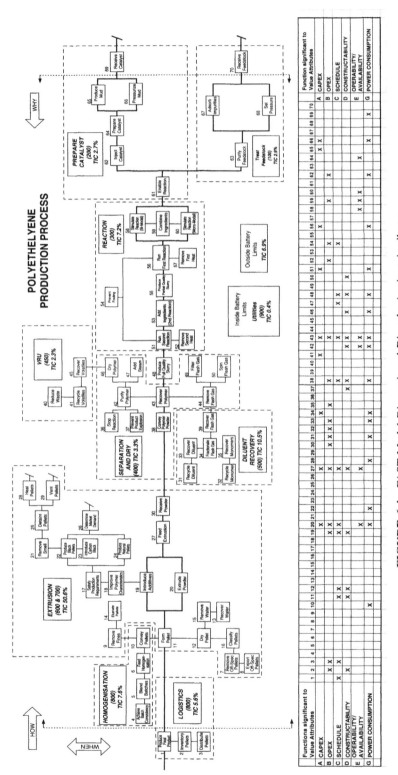

FIGURE 6.4 FAST Model Dimensioned Directly Against Attributes

Measurements give managers insight into the effectiveness of the way that solutions perform underlying functions. This is stated now so that a clear view of the collective thinking is prominent in our minds. In a moment we will not look to functions first but to phenomena that we can measure and so change our perspective while keeping in mind a relationship between functions and a solution's measurement. That is, so far in the book we've started by building a FAST model and then correlating it to the performance of solutions so that we are in a position to swap old ways of doing things for new ways of doing things. Next we start with the logic of the "old way" in order to drill down into the underlying functions and open the way to innovation. The reason we are doing so is to allow you to find out why we model one set of functions rather than another. It's about creating a clear line of sight from strategic value through issues of concern to the underlying functionality. The FAST model is in line with the way we choose to innovate.

When a client phones us to ask for assistance, we need to establish the purpose for the call. We don't automatically accept what we are told and suspect that at this initial stage only a superficial view of the issues of concern has surfaced. We may be listening to symptoms rather than the underlying causes of those symptoms. We must try to establish who the decision makers are and how they will be judged as to having made good decisions. Often, this involves determining the management level and decision authority of the person making the initial phone call. We have to start building a view of which stakeholders are involved and how they will define value. Let us now run through an example and in so doing make the links among reality, issues of concern, and the FAST model explicit.

Imagine a situation where a manager is facing falling profitability but does not know how to turn the situation around. We were told that retained profit was growing but not as quickly as management wanted. It did pick up for awhile but has since taken a nose dive. We begin by building a story of what has happened. For example, senior management believed that they were overresourced with engineers and had strategic ambitions to increase retained profits by reducing engineering spending in order to get some good financial results before the annual general meeting with shareholders. Their plan was to reduce the engineering budgets for staffing.

From the narrative above we can pull out some variables and how the trends were anticipated.

- Engineering spending is higher than it needs to be.
- If engineering budgets are cut, costs will fall.
- If the number of engineers is reduced, the company will still be able to meet workload demands.
- If revenue is kept as is and costs are reduced, retained profit will increase.

Here we see some management theories being made explicit. At this stage we are not judging them. We are simply getting them out into the open. Figure 6.5 borrows from Peter Senge's work[2] and system models. It shows an implied causal relationship between the variables. The effect that one variable has on another is shown against time. The solid lines represent actual trends, and the dashed lines represent perceived trends. So as engineering spending is lowered, retained profit continues to grow—at least that's their theory. One variable causes another to change, as indicated by an arrow. These variables are examples of what we already discussed when dimensioning

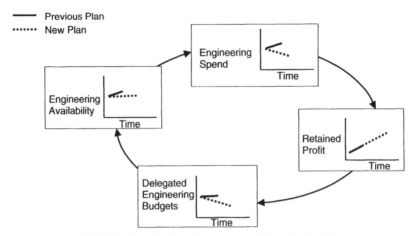

FIGURE 6.5 Making Management Theories Explicit

a FAST model. As a "thing" functions, it is assessed by way of performance measures. As a car's engine functions, we could measure temperature, torque, fuel consumption, and so on. As an organization functions, we can measure its effectiveness in terms of its design, staff turnover, skill levels, and profitability. The system dynamic approach models the results of solutions performing functions, and that's why we can use this method to glimpse higher-level functioning, especially when we don't have enough time to make the functioning as explicit as we would like. Although the system dynamics literature does not refer to functioning, the relationship between functions and the effects that result and are measured have been implied throughout this book and are now made explicit.

As we listen to the stories being told, we ask questions to tease out other theories. In Figure 6.6 we acknowledge that a theory exists about the perceived engineering workload (see the new graph in Figure 6.6 that was not in Figure 6.5). The perceived engineering workload is represented by a solid line on the graph that suggests a management theory that the workload is currently falling. It is also necessary to

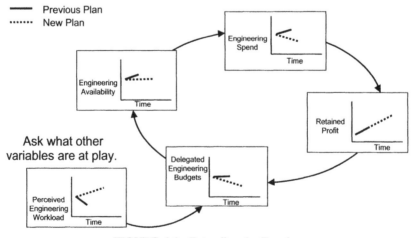

FIGURE 6.6 Extending the Enquiry

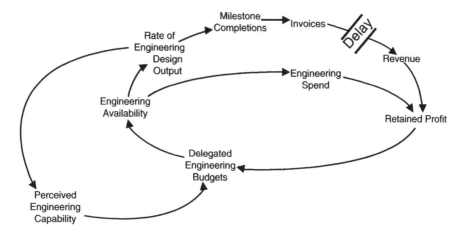

FIGURE 6.7   Making the Decision-Making Complexity Visible

remember that we are mapping theories between data sets and are implying causal relationships. This implied meaning has credibility only if there really is a causal relationship, and as with the FAST model, the truth value of the model is a reflection of the accuracy that it purports to represent. A prudent FAST modeler would also seek out evidence to prove that the management theories are correct, for example, by speaking to the engineers and by looking at key outputs to see whether management theories are as correct as assumed.

From such an investigation we glean other causal theories of the business. For example, let's say the principal engineering manager said something like: "The number of engineers affects the rate of design output. Having more engineers, we can then release more designs in a given time period. This is important to the firm, as we get paid on the completion of milestones in the master schedule. Once we hit the milestone, we can send an invoice to the client, and after a few weeks we get paid." We can then add more to our system model as shown in Figure 6.7. Here we've added the variables in the foregoing statement to a simplified version of Figure 6.6. What we are doing is making the complexity visible and in so doing enabling a clearer view of what is going on. But that's not all. We are also allowing a dynamic representation to unfold, so next we make the statistical correlation trends explicit. Remember that if the relationship between two variables has a positive correlation, then as x rises, so does y. Similarly, if the two variables have a positive correlation, then as x falls, so does y. Positive correlation means that the trends of the two variables move in the same direction if indeed there is a true causal relationship between them. The flip side is negative correlation. With negative correlation the two variables move in opposite directions. As x rises, y falls, and as x falls, y rises. Look at Figure 6.8 and work your way through each variable to check if you agree with the causal relationships it suggests; for example, if *Delegated Engineering Budgets* falls, would we agree that *Engineering Availability* would fall as a consequence?

We can now build a story that we can test among the managers. For example:

1. It would seem that retained profit was growing but not as rapidly as senior management wanted.

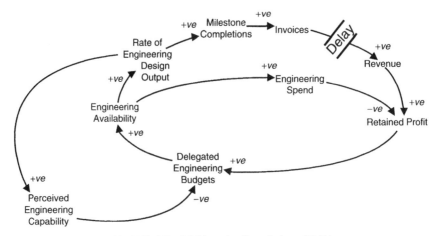

FIGURE 6.8 Making the Correlations Visible

2. Someone made an inference that there were more engineers than were actually needed for current workloads.

3. Management decided to reduce the engineering budget, as it was perceived as being too high for the then-current workload.

4. Engineers were laid off, and thus a salary saving was achieved.

5. Because revenue was arriving as a result of past work and because of the delay in the invoicing and payment processes, the reduced engineering spend today meant that monthly profits rose.

6. Even though sales increased, the engineers were expected to work harder and therefore increase their productivity.

7. New work and lower engineering availability meant that the rate of engineering design output began to fall.

8. Management realized that there was a shortage of engineers, but as retained profits were rising, they did not intervene; it looked as though they had got things right.

9. As engineers faced tougher working days and saw little chance of this changing, their morale began to plummet and some even started looking for jobs elsewhere.

10. The lowered rate of design output also meant that fewer milestones were being achieved as rapidly as before.

11. The lowered numbers of milestone achievements meant that fewer invoices were sent to the clients.

12. The lowered number of invoices caused revenue to fall.

13. Management currently believes that the only way they can return to profitability is to lower staffing costs again.

14. Someone brought us in to help create more value, and we began to search beneath the symptoms.

By building such representations we can quickly see the systemic nature of the issues facing management. The solutions to organizational functioning were changed, which then led to a different level of performance. Think about an internal combustion engine that functions with petrol being used to force the pistons to move and what would happen if inferior petrol with a low ignition threshold were used or if someone were to put diesel fuel in by mistake. Many of the functional relationships remain, but the changed solution brings a change in the way the functions are performed and thus the way the engine performs. It is the same with organizations, and sometimes the change can occur outside the company, such as changes in level of taxation, which again affect performance.

Following such conversations with managers, we then ask three questions to make management issues explicit and to identify which attributes we as managers can change. Next we explain these three questions. In Figure 6.9 we show that in this example the only things that management can easily change are (1) the delegated engineering budgets, (2) engineering availability, (3) the rate of engineering design output, and (4), milestone completions.

Remembering that the strategic goal is greater profitability, we can now engage in a line of inquiry that would help us to know whether we are improving or degrading capability. Are we swapping ways to perform functions in the FAST model that actually lead to improvement, or not? It is at this point that we would ask three questions:

1. What problem or opportunity are we about to address?
2. Why has this problem or opportunity arisen?
3. What would happen if we did nothing?

In asking these questions, we might search for:

- Modes of hiring engineers, such as employing them directly or contracting them in
- Modes of increasing engineering availability, such as using the Internet to hook up with engineers in other parts of the company, either home or abroad
- Modes of accelerating design output, such as allocating tasks to engineers based on previous experience or using past designs for tasks that do not require major innovation (i.e., reducing the short-term need to invest in learning curves)

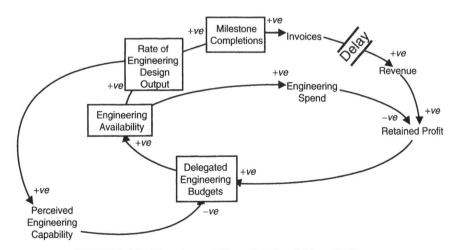

FIGURE 6.9   Focusing on Where We Can Make a Difference

- Modes of distinguishing between milestones so that short-term cash flow can be boosted

From such conversations we articulate attributes that managers can choose to improve or degrade. An important but subtle point is being made here. We do not ask managers to determine which attributes are important. We ask them to identify which, given today's situation, need to be improved at the expense of others. Our weighting focuses on the relationship between those things we can change as managers and how such changes affect the strategic value we aim for.

It is now possible to say that of all the things we could depict in a FAST model, in this example the FAST model should focus on how engineering capability is managed. We have brought a focus to our study because we have developed a systemic understanding of the management's issues of concern and that these are grounded in evidence rather than opinion. If we can identify the functions and map on to them how things are done, currently we are in a position to consider alternative solutions and in so doing open the way to innovation. We now show a case study for construction management students to see how FAST models enable a rich learning within an innovation process.

CONSTRUCTION MANAGEMENT CASE STUDY

So far we have explained the relationship between bounded attributes in a range of goodness and how they are used to assess the effectiveness of new solutions that replace old ones. The FAST model provides a coordination role in this process where we swap different ways of performing functions and assess them via attributes. A key problem is identifying those attributes that if changed would increase value. We could simply pluck some out of the air, but that would lack the systematic enquiry we have argued throughout this book; such attempts often pick out symptoms and fail to tackle the underlying causes of lost value. Let us look at a situation in which the main focus was in identifying attributes and setting in place lines of thinking that direct the FAST model and the selection of knowledge held by team members.

By making management theories explicit, we gain a perspective that is not commonly shared. Different people will believe that the key attributes are not what others claim. How can we develop a shared identification of those key attributes? We do so by converting variables from a system dynamic model into attributes and then identifying the key ones that we as managers can influence. The selection of key attributes allows us to align those functions whose solutions would lead to improvement if swapped for better alternatives; this again shows the link between attributes and FAST models. As we link sources of data (i.e., evidence) to theories, we need to reduce production volume. We not only understand the systemic relationships that organizations work within, but often how their strategies work. In the example above (see Figure 6.8), adapted from an episode in North America, the mistakes the management team made could be seen as embarrassing to them, as they could not see the contradiction in their strategy that sought to "shrink its way to growth." In this case study we explore an educational episode. Students studying construction management were challenged to become a one-stop-shop property development company. They were to explore how to build a housing scheme on a brown field site that was previously the site of a steel factory. The key task, presented as a university student project, was to identify attributes and set themselves up to construct a FAST model that allowed them to swap proposed solutions for better ones.

The key issues were:

- *Context*: The students are to set up a hypothetical medium-sized development and construction company and improve value in a proposed development scheme.
- *Problem*: They are overwhelmed by complexity and do not know how to engage with a design team to achieve the best outcome. As such, they have retreated to the comfortable thinking associated with their training to date.
- *Constraints*: Their company is to build 30 houses within 12 months and for a cost no greater than £60,000 per unit on a brown field site.
- *Learning objective*: to expose the students to strategic thinking.

The students were broken into small teams and then allowed to develop their own concept designs for a specific site. As can be anticipated, they laid out their traditional houses in a traditional format without considering functionality. That is, they rushed into preconceived solutions. For them the issue was about how best to apply their training, which leaned to the practical side of building; in other words, the delivery of preconceived solutions in an efficient way took precedence, with little regard for effectiveness. A FAST model would have forced them to think about different solutions, and the range of goodness and attributes would force them to think about value.

We began by explaining an overview of the value engineering methodology (see Figure 6.10). We start by planning what we want to achieve in the pre-event phase, which we've already discussed in this chapter as strategic problem and opportunity framing (stage 1 Figure 6.10). In this stage we seek to understand the systemic context and draw out attributes that we as managers can change, and then weight them with respect to which would give us the biggest value improvement. Then we have to figure out how things are currently done and how we can change that. This is where the FAST model comes in (stage 2). After we have built the FAST model, we dimension it by

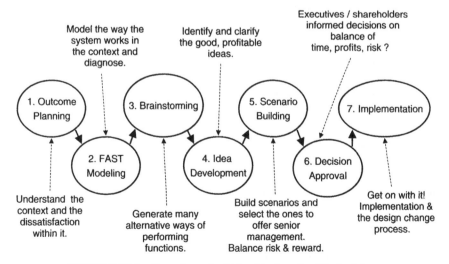

FIGURE 6.10 Overview of a Value Engineering Methodology

cross-referencing the "measurable" from existing solutions back to the functions they perform. We then brainstorm (stage 3) to generate many alternatives to the way that functions can be performed and select the ones with the best prospects to receive some good-quality research and development time (stage 4). Once we have some trustable ideas in which uncertainties have been translated into risks, we set about combining them to form synergies that become scenarios (stage 5). We then offer a collection of these combinations (i.e., scenarios) to management for their approval (stage 6). Management then makes a decision and commits resources accordingly, and it is following this stage that we begin value tracking during implementation to make sure that what was planned happens, and if it does not, that lessons are captured and learned (stage 7).

At this stage the students seemed to be very comfortable and in the discussions did not show any awareness that by thinking differently they could unlock innovation and create more value. They had not realized that the way they were thinking was a process and could be structured and improved. We next engaged them in some basic questions:

- Why would a development and construction company be necessary in society?
- What is the reason for its existence?
- What is the relationship between why the company exists and an ability to generate profit?
- Which stakeholders affect the company's ability to make a profit?
- Why would customers give them money?

This led to the following starting point:

1. *How does money flow into and through the value chains?* People buy houses with mortgages, and this money passes through the property developer to the design, management, and construction teams that build.

2. *Who directs revenues?* The customer, either by giving money for the purchase of "your" house or by selecting a rival offering.

3. *Who incurs cost?* The firm does. They are second-tier customers who pay designers, suppliers, construction personnel, and managers to create the product (e.g., a house).

4. *Who guides the rules, and what is their interest?* The government, legislature, local authorities, and professional bodies. They look after national interests (e.g., energy consumption and building regulations) and local interests.

What we were doing was engaging them in a conversation that resided outside their normal focus on the practicalities of building houses. The students were talking about the relationships between strategic variables, as we discussed earlier in the chapter when discussing the pre-event. We were starting to get them to make their theories explicit so that they could reflect on them and check whether or not they were valid. As we undertook the conversations, we got them to talk about variables rather than events, and this let us explore profitability in terms of cost drivers and revenue drivers. Without them being expressly aware, the students were actually starting to talk about attributes and were starting to develop a better understanding of success and what potential additional value was possible.

We then explained what we meant by causal relationships and correlation. To make things easier for them, we showed positive correlation as "S," meaning that the variables move in the same direction, and "D," for a negative correlation. Then we began a brief discussion about macroeconomics, which opened the door to discuss the property market in general before honing in on the particulars of their proposals (see Figure 6.11). The students were articulating their theories, which combined their project's internal and external environments. We also talked about lagging effects that are not shown on the figures. They were getting a richer understanding of the issues their decisions would have to deal with. The discussions here brought in some of their learning from courses (e.g., in economics) and helped them to start seeing how such learning might fit with the practicalities of construction. In Figure 6.12 we made a link between macroeconomics and affordability.

Next we started to discuss how the economy and affordability might influence people to want to live on their scheme, so we were elaborating a theory of demand (see Figures 6.13 and 6.14). As the conversations took place, the students were engaged in the act of theory mapping and sense making. They were beginning to realize which key

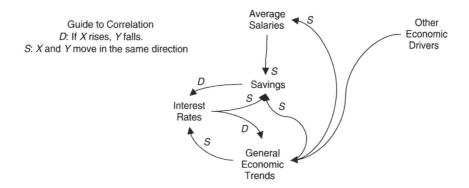

So if the economy is growing rapidly, that, combined with other economic drivers such as inflation and savings, will influence the Federal Reserve to raise interest rates, and if they do raise interest rates, the economy would slow down after a lagging effect completes. What we are showing is the relationship between variables that later influence our choice of attributes to use in the assessment of proposed solutions to funtions in the FAST model.

FIGURE 6.11   Role of the Economy in the Ability of People to Buy Houses

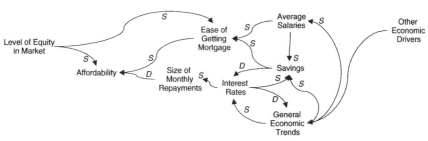

It's about building the big picture with key team members so they develop a shared understanding.

FIGURE 6.12   Impact of the Economy on Affordability to Buy

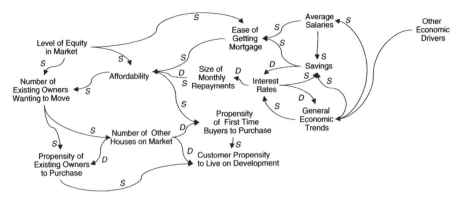

FIGURE 6.13 Impact of the Economy and Affordability on the Commercial Feasibility of the Development Planned

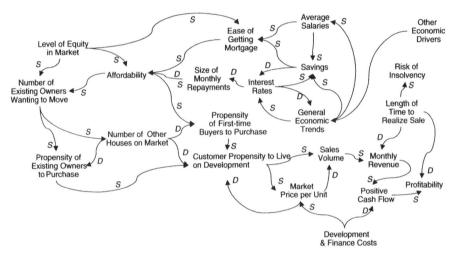

FIGURE 6.14 Dependency of Commercial Feasibility on Generating More Revenue Than Cost Expended

attributes and what kinds of functions needed to be in a FAST model. We then brought the relationship of microeconomics into the story, in particular how their schemes might fit with this. It was about this point that we brought in the concepts that they were familiar with and started to discuss the practicalities of building "on time, within specs, and within budget." At this stage we talked about the quality of the building and the management of the building process and started to question whether there really was a causal relationship between cost and quality (see Figure 6.15). This led to an interesting discussion, which allowed a view of what types of solutions to key functions such as *Build Home* would be inappropriate. Some questioning from the students searched below their existing knowledge base and into some fundamental questions, such as the relationship between cost and ethics. They were starting to understand how to recognize better solutions and the need to model certain types of functions in their FAST model. It was from this episode that a fundamental question was asked: Can design value and construction value create a desirable place that satisfies Mark Twain's "location,

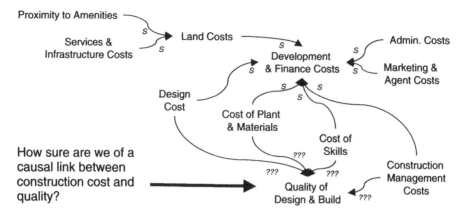

FIGURE 6.15   Key Construction Drivers

Location, location." What functions would have to be performed to achieve this? What attributes would allow us to know whether the ways chosen to perform functions in a FAST model would result in more value? These were the types of questions that arose.

The students were already realizing that to add value they needed a multidisciplinary team that united design and production. They needed a method to achieve this and were discovering the need for FAST models. This was their realization. It is our theory that to "discover" brings a deeper level of learning that is much richer than simply being told by someone that it's better to unite design and production. We placed the bits together and began to inquire about the role of design. It was no longer an architect telling them what to do. It was about collaborating to add value (see Figure 6.16). They realized that people don't buy cost; they buy things that they presume will benefit them. They buy value in some form of improvement to their lives, which is a solution that performs a valued function. They pay money as they make trade-offs between preferences and resources. This was a marvelous realization, for it opened the door to attributes that we as managers can change. In Figure 6.17 we show the completed model that made all the management theories both explicit and linked and now have an ability to consider the complexity of the situation. If we had started with this model, few would have had the enthusiasm to figure out what it was revealing. By engaging the students, we built a series of insights that underpin both the pre-event and the focus of the FAST model. This motivates us to ask: "What can you as construction managers change in all this? What do you need to influence if you do not control it?"

We had now engaged them in strategic thinking. It was no longer about cramming as many units on a plot and building them as quickly and as cheaply as possible. It was about thinking how to increase value through the choice of solutions to perform functions. Some of the questions we asked at this stage were:

- How do construction managers interact with or influence marketing departments or customers to ensure that they build to quality specs that match the customer's definition of quality?
- How do construction managers interact with or influence local government agencies to get as much value out of planning and building regulations as possible for the customer?

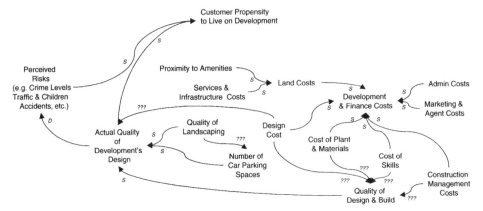

FIGURE 6.16   Role of Good Design

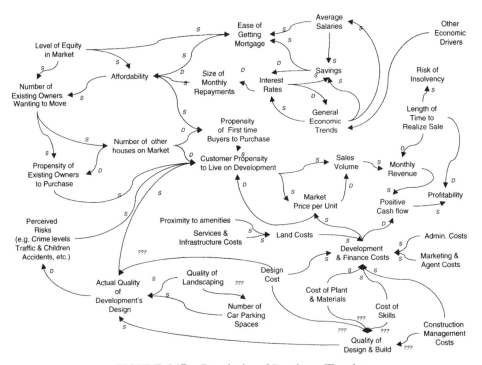

FIGURE 6.17   Complexity of Dominant Theories

• How do construction managers interact with or influence suppliers to ensure customer value?

Figure 6.18 shows which variables the students selected as those they could influence and by so doing influence the way the system performs. They were selecting the
attributes for the range of goodness. These variables were then treated as attributes.
By attributes we mean that they can take on a specific value (i.e., number) from a
range of possibilities. To enable the remainder of the process, we selected one group
of students and looked at their proposed scheme. We next established the upper and

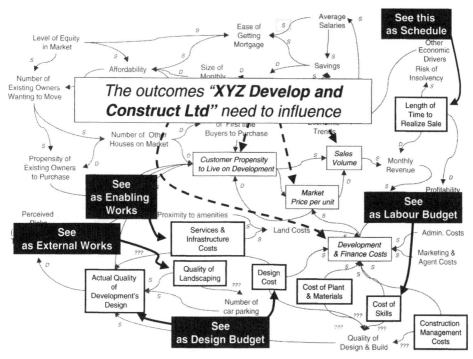

FIGURE 6.18 Variables that Construction Management Can Control as Attributes

What level of performance would be excellent or *innovative (10)* ?

Where are we now (⋆)?   Beyond best-in-class

Mapping the Base Case	1	2	3	4	5	6	7	8	9	10
A. Schedule (Months)	24			18 ⋆						14
B. Labor Budget (£000)	240			205⋆						180
C. External Works (£000)	58 ⋆			48						38
D. Enabling Works (£ M)	£1M ⋆			£0.7M						£0.3M
E. Design Budget (£000)	130							80 ⋆		70
F. CM Costs (£000)	200					160 ⋆				120
G. Plant & Material (£000)	160⋆		100		45					50

FIGURE 6.19 Setting Boundary Conditions on the Attributes

lower boundaries for each attribute as shown in Figure 6.19. Column 1 represents the lowest level of acceptance that management would tolerate. If the proposal could not prove that it would meet all of these lower boundaries, it would not win funding. Column 9 represents what we would imagine the best companies to be able to achieve, and therefore column 10 is our attempt at defining what is better than best in class. The asterisks locate where this student's project was at that stage. So they now had a means to distinguish a great from a poor collection of solutions and knew they needed a FAST model to allow them to manage the swapping of solutions.

Customers do not necessarily value all attributes equally, so when considering which solutions to perform functions in the FAST model we need to keep focus on customer preferences. In Figure 6.20 we weight the attributes using a paired comparison approach. In this technique we seek to identify which attribute, if improved, would yield the best value. It is not asking which is most important but which one needs to be improved more than others.

In Figure 6.21 we calculate a value score by multiplying the weighting from the paired comparison (see Figure 6.20 and row A in Figure 6.21) by the column number of the range of goodness (see Figure 6.19). In an ideal world each weighting would be multiplied by a position that was better than best in class and so would be the weighting multiplied by 10 (see row B). The reality is the weighting multiplied by the column with the asterisk in it to mark where the students were at that stage (see row C). The value score (see row D) provides us with a datum that we can use to check whether or not we are really adding value. However, now we know what the value score is, but we still don't know what functions need to be performed to either increase or decrease the score. Remember that when a particular solution may have a cost or generate revenue, it is also performing an underlying function that will be shown on the FAST model.

Figure 6.22 shows the *how–why* logic path, which makes the intention explicit. Here we see that the basic function is *Increase Customer Value* and the purpose or higher-order function is *Attract Buyers*. The students were no longer thinking about construction schedules or materials. They were thinking about the functions that will make their scheme successful.

In Figure 6.23 we make the causal logic explicit and here use *if–then* rather than *when* to show how this can also be used. If we *Ensure Construction Quality*, we *Increase Aesthetic Appeal*. If we *Increase Aesthetic Appeal*, we *Communicate Design Excellence*. The functions are sequentially dependent on the *if–then* rule. There may

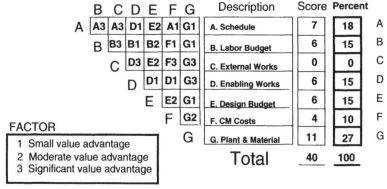

Which levels will deliver the greatest value based on today's plans?

	B	C	D	E	F	G	Description	Score	Percent	
A	A3	A3	D1	E2	A1	G1	A. Schedule	7	18	A
B	B3	B1	B2	F1	G1		B. Labor Budget	6	15	B
C		D3	E2	F3	G3		C. External Works	0	0	C
D			D1	D1	G3		D. Enabling Works	6	15	D
E				E2	G1		E. Design Budget	6	15	E
F					G2		F. CM Costs	4	10	F
G							G. Plant & Material	11	27	G
							Total	40	100	

FACTOR
1 Small value advantage
2 Moderate value advantage
3 Significant value advantage

Zero means that External Works ranked seventh out of seven in order of need to improve, NOT that it has no value.

FIGURE 6.20 Weighting the Attributes

❏ Define the base case value. Added value must be above this datum!

weighted score = position of base case on attribute x weighting

Score: 354
Target: 650 minimum

	ITEM	A	B	C	D	E	F	G	TOTAL
A	Weight	18	15	0	15	15	10	27	100
B	Ideal Value Score	180	150	0	150	150	100	270	1,000
C	Current Position	4	4	1	1	8	6	1	
D	Current Value Score	72	60	0	15	120	60	27	354

FIGURE 6.21   Calculating a Value Score

Understand clearly what functions need to be performed.

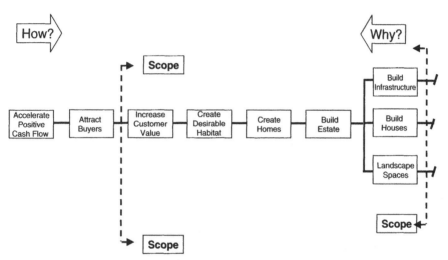

FIGURE 6.22   Making the Intentionality Explicit

be more, but that's not the point in this educational study. What we are more interested in is revealing a method that teases out and structures the various types of thinking that go into a successful project. The FAST model was then dimensioned with respect to the attributes and various types of cost. This enabled the students to identify which functions had solutions that needed to be swapped. Then they followed the procedures already discussed. They developed ideas, combined them into risk–reward scenarios, and selected the ones that would be offered to senior management for budgetary sanctioning.

What this case study has done is make the link between the pre-event and the FAST model explicit. It has shown how theories are used to link variables and then how those variables the team can influence lead to the articulation of attributes that address issues of concern. Following this, the attributes are weighted with respect to how, if changed, they would affect the strategic value. It is from this process that we establish a value score as a datum and become aware of which current solutions need to have their underlying functionality made explicit.

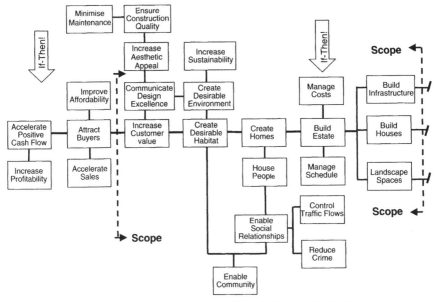

FIGURE 6.23 Making the Causal Theories Explicit

The better we understand the problems and issues, the more chance we have of solving them adequately. Next we explore a case study that examines this relationship between the strategic context and the FAST model in more detail.

INFLUENCE OF ATTRIBUTES AND INCENTIVES ON FAST MODELING

Have you ever walked into a computer store to buy the latest "must have" computer software program and found that the product you wanted sat next to another attractive product? The problem is that the store stocks these two competing products, one costing $150 and the other $500. You examine both products. They look almost identical, with the exception of a few cosmetic differences. Reading the specifications reveals that they are also about equal in performance and complexity. If performance is a higher-valued attribute than price, which product is the better choice? You inquire whether the more expensive product has inherently better quality, aftermarket services, and is more backwardly compatible and update friendly. Is that the reason for the price difference, or is the high-priced producer new to the market, thus lacking the manufacturing experience and capabilities necessary to produce products cost-effectively? The following case study explores how incentive schemes can be used to ensure that customer value is maximized through FAST modeling.

SOFTWARE ACQUISITION CASE STUDY

The case study we are about to describe is similar to the story above, but on a much larger scale. The problem and challenge of procuring leading-edge technology is in assessing the responses to the *invitation to bid* (ITB) process, where there is little past history and data to guide selection of the contractor. This case study describes such a condition and how the FAST modeling process helped resolve some very challenging issues. The project involved a government agency's need to develop new system software for military use. At issue is creating a way for personnel to recognize and distinguish quickly between friendly and unfriendly combatants in the field. That choice, to engage or to abandon the military mission based on the software's performance, is literally a life or death decision.

During the preliminary *request for proposal* (RFP) phase, three software producers were selected based on their apparent understanding of the performance requirements and their proposed software development approach. However, the range of bids submitted by the competing software providers was spanned over 350 percent. The agency appreciated that this project highlighted bigger problems inherent in the way that software was specified and purchased. They employed a value engineering process, specifically a FAST model, to focus on the broader issue of how to acquire software products effectively. Reviewing the history of similar problems, the agency determined that it needed a way to assess requirements for acquiring software products. The software products would need to conform to critical performance attributes within a reasonably short development cycle, perform to required specifications, and be easily maintained in the field and upgraded when necessary. The software product described in the case study was used as a focal point in assessing and developing a new software acquisition process. So this case study is about improving an organizational procurement capability.

In addition to the agency's engineering, contract, and program management staff, other users and affected organizations were invited to join the study team. The contracting software providers were also invited and participated as team members. Initially, the potential contractors were hesitant to participate because they did not want to divulge their company's sensitive information to competitors. It was explained that their experience was valuable and needed to create a better procurement process and a level field that would preclude inadvertently giving one contractor an unjustifiable advantage. They agreed and participated actively in the study as team members.

The FAST team determined that the development and integration of software products and related services is not only labor intensive but that there exists a growing need for stronger intellectual property protection for the vendors. The issue of how to handle intellectual property was addressed separately, outside the scope of the study. The principal project issues were described as how the agency can clearly specify and obtain, under guarantee (or warranty), the required technical performance and maintenance of acquired software products, at a fair price. It was also determined that proper contractor incentives should be included in the agreement as a way to stimulate creative efforts that go beyond critical performance requirements to deliver a high-performing product at a lower price and lower maintenance cost.

Prior to beginning the FAST model, the team was instructed on how to create a FAST model describing the most "ideal" process to achieve the project objectives, disregarding development cost, organization boundaries, governmental constraints, and conflicting or restrictive directives affecting the project. This approach to FAST model development is a way for a FAST team to classify the assignment as a "clean sheet," or a new concept opportunity, rather than reverting to patching or modifying the current system. They chose the former approach.

With a blank piece of paper in front of you, it is one thing to say "disregard the world about you" but quite another to overcome the mental inertia that keeps bringing team members back to the real world. However, this attitude of no constraints must prevail because it encourages a far greater number of ideas, the quality of which can be assessed at a later stage. The prospective outcome is improved as good new ideas are refined and brought back to the reality of the project environment.

Figure 6.24 is the FAST model produced by the FAST team that displays functionally the approach to a new software procurement process. The FAST model shown is the result of the collaborative effort of participants with different professional disciplines and both technical and practical knowledge. This wider stakeholder inclusion represented the various agencies that would be affected by the outcome of the study. Also represented were the software developers or contractors who would be called on to produce the software product in compliance with the technical and contractual requirements of the project.

The sensitivity matrix shown in Figure 6.25 is another example of a modified RASI, described in Chapter 5 (see Figures 5.2 and 5.3). The matrix is presented as a separate figure so that each part of the FAST model can be expanded to improve legibility. Normally, the RASI matrix and FAST models are combined to facilitate team analysis. The matrix in this figure identifies those agencies and the software provider who were affected by the procurement process, many of which were team members and participated in constructing the FAST model. The FAST team representatives were also involved in some way with the implementation and performance of the product and processes developed to implement solutions to the FAST model. Additionally, the matrix identifies those responsible and those in support of or "moved to action" by a particular function in the FAST model.

The sequence of first producing a FAST model, then deciding responsibilities and accountabilities is important. By instructing the team that the FAST model must show the functions of the organization without considering who will be responsible for those functions, territorial disputes are set aside. When the FAST model is complete and the participants agree to its validity, the task of building the RASI matrix can begin. At this time territorial disputes often surface again. However, all have agreed that use of a FAST model was feasible and represented their discipline's involvement. The key issues now concerned organizational boundaries, not process validity. Such territorial boundaries and their implications bring a strong motivation to resolve which organizations and key personnel are responsible for selected functions and which are in support roles. Here we use a FAST model to check organizational designs and whether roles and responsibilities are appropriately staffed.

If the FAST model-building process was reversed, that is, the roles of the representative team disciplines were determined first, then the FAST model constructed, the result would be a poorly crafted process replete with territorial compromises that would raise issues during implementation.

VALIDITY OF A FAST MODEL

In earlier chapters we learned that most FAST models are case specific. It is not useful to judge a FAST model as right or wrong, but it is useful to consider whether or not the FAST model is a valid representation of the project issues to be resolved. Is it felt to be an accurate portrayal of the functioning? The FAST model is judged valid by achieving

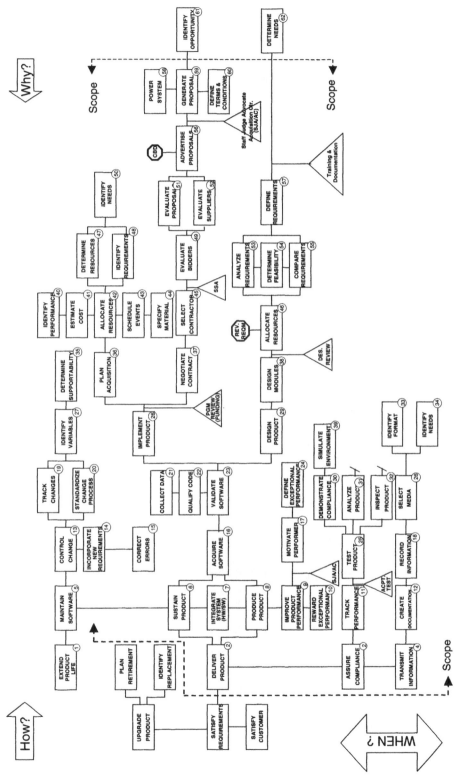

FIGURE 6.24 Software Acquisition Process

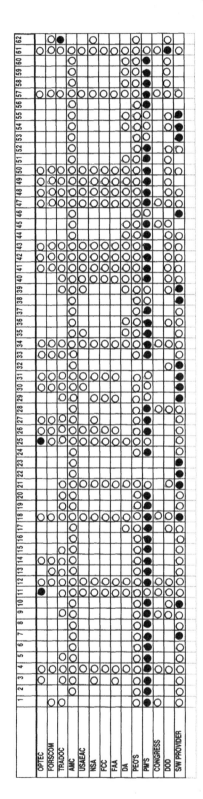

Indicates those responsible

Indicates those in support of, or moved to action

FIGURE 6.25 Software Acquisition Process: RASI

142

team consensus from a group of knowledgeable experts sharing theories within and across their professional knowledge domains and seeking corroborating evidence from outside the FAST modeling team. That is, if the members of the FAST team, who represent different disciplines and interests, all agree that from their perspective the FAST model adequately represents their theories about what is functionally necessary, consensus is achieved. It is inevitable that when viewed by sources outside the FAST team that additions or corrections to the FAST model are offered. A major reason is that the viewing of a FAST model by someone not involved in its construction brings a need for sense making which lacks the rich discussion that went on while it was being built by a multidisciplinary team. If an external expert makes a valid suggestion, it should be included on the basis that it increases the accuracy and adequacy of the FAST model; if it is simply a matter of semantics, the effects on the model's usefulness should steer the decision as to whether to amend or leave things as they are. Selecting who will and who will not serve on the FAST team is therefore an important consideration because the disciplines they represent and their professional level and competence relate directly to the quality and credibility of a project's outcome.

Pre-event's Role in FAST Modeling

Among the issues resolved during the pre-event or framing of a project is how a FAST model will be constructed and what disciplines need to be represented while constructing the model. Addressing the first issue affects the way the FAST model is developed. Key concerns are defining the problem or opportunity and project goals that if achieved would resolve the project issues; identifying, defining, and ranking the major project attributes; and surfacing project constraints. Resolution of the pre-event topics will determine the following FAST issues:

- Should the model also describe an improvement or a radical innovation?
- If we seek a radical innovation, is it within existing or new technologies?
- What should be the model's operating level of abstraction?
- What is the scope (e.g., the study boundaries) of the FAST model?
- Should the model identify major project constraints, negative functions, or unwanted functions?
- How should the model identify the project's approval authority?
- Considering the project goals, how should the model be dimensioned, and which metrics should be selected to define the dimensions?

The answers to the foregoing issues will also begin to form the credentials of the FAST team participants. Collectively, the FAST team should represent three points of view that form the basis of a business case. The dominant perspectives found in the answer to the three questions that follow also govern the team makeup for any problem-solving assignment. (We should, however, check that those perspectives are correct by corroborating against external evidence.)

- Who owns the problem (or opportunity)?
- Who is responsible for solving the problem (or resolving the project issues)?
- Who will be affected by the resolution of the problem (or the resolution of the project issues)?

Areas Defined by a Scope Line

The search for answers to the three questions above can be found in the general structure of a FAST model. The scope lines form the boundaries separating the three areas. Functions between the left and right scope lines are within the boundaries that define the project under study. Members of the FAST team who represent particular areas must have the competence and managerial authority to address the second question: Who is responsible for solving the problem?

In the area to the left of the left scope line are the higher-order functions. These functions justify the existence of the basic function, as they provide a sense of the purpose found by asking the *why* question. The performing functions that reside between the scope lines can affect the quality of the product (the basic function) that supports the customer or the recipient of the product (higher-order function). Higher-order functions are "independent" of the functions within the scope lines. They are therefore purposes or outcomes used to judge the performance of the basic function, but they can't change those functions unless they themselves change why the basic function exists. This can offer clues as to who is concerned enough about the problem to request a resolution. It will also answer the first question: "Who owns the problem?"

The area to the right of the right scope line houses the dependent functions. Since functions to the right of another function depend on the function to its left, changes in the way functions are performed will affect the performance of those dependent functions. It will also help to answer the question "Who will be affected by resolution of the problem?"

Reviewing the three areas in the case study example (see Figure 6.24 and the scope lines that divide the three areas), it appears that the customer in *Satisfy Customer* are the agencies representing the various military commands. Technical and contractual members of the procuring agency support the area between the scope lines. The software providers or contractors respond to the dependent functions *Identify Task* and *Determine Needs* in the area to the right of the right scope line. Organizations representing the functions described in the three FAST model areas, within and either side of the scope lines, are shown in Figure 6.25 (e.g., OPTEC, FORSCOM). Not all of the organizations identified on the modified RASI need to be physically present on the FAST team. The U.S. Congress, although not serving on the FAST team, is represented because their various committees perform oversight functions and are responsible for approving the budget needed to develop, install, and maintain the software product. The expectations of these distant stakeholders must be included in conversations needed to develop the FAST model.

Resolving the Incentive Issue

Returning to the data created during the pre-event phase, the FAST team discussed the possibility of using as the basis for creating performance incentives the project's scale of goodness chart, showing the software project attributes selected (see Figure 6.26). The team determined that the attribute values in column 5 would represent the target performance values. Extrapolating the values in column 10 represented outstanding stretch goals just beyond the horizon of expectations. Conversely, the values in column 1, although reluctantly acceptable, would represent a penalty for an "incentivized" contractor. Not all attributes from columns 1 to 10 form a straight-line series of values. Attribute performance curves can be linear, exponential, or a step function. Filling in

ATTRIBUTES	1	2	3	4	5	6	7	8	9	10
Response time	1 sec.		.75 sec		.5 sec			.35 sec		.25 sec
Innovation	Full Mil Spec				Mil/Com Specs					Full Com Spec
Friendly Identification	95.0%				97.0%					99.9%
Supportability	Mil/ Ada				Ada95 w/com bind.					Full com/ISO
Program cost	125%				100%					70%
Flexibility	Host Dedicate				DoD Platform					NATO Platform
# of simultaneous response	3	3.5	4	5	6	7	9	12	15	18
Quality (MTBF hrs.)	2760		2933		3105			3278		3450

FIGURE 6.26 Software Acquisition Process: Scale of Goodness

more of the attribute values on the scale of goodness chart will display the curve's shape as various characteristics along the attribute are plotted.

Now let's suppose that during product development an engineer finds that he or she can improve significantly the target performance of one attribute, say *Response Time*. In doing so, however, the engineer would need to reduce the performance of *Friendly Identification*. The engineer approaches the program manager with this performance trade-off. The program manager responds, "No, you will not do anything that will reduce the performance of *Friendly Identification*." In accepting the program manager's decision, the engineer thinks, "So, all the attributes are not equal. Some are more important than others. Wouldn't it be nice to know the relative importance of the attributes before starting product development?"

The purpose of a paired comparison is to determine an attribute's relative importance to value improvement. In the case study, each of the attributes was assigned a relative range of importance, the result of evaluating the attributes by a paired comparison (see Figure 6.27) performed by the case study team. As stated previously, the process of paired comparison will often result in dropping out an attribute that scored zero. In this case, the team elected to leave that attribute in the system, because *Innovation*, the attribute dropped, is important even if, as in this case, it ranked last among all the attributes. Innovation was kept in the system by changing the attribute's value from zero to 1 and raising the score of all the other attributes by 1.

The relative rank of the attributes and their weight (see Figure 6.27) were incorporated in the performance requirements section of the request for proposals, allowing competing contractors to use the attributes and incentive schedule to determine their bid strategies. Figure 6.28 is a scale of goodness chart, modified to demonstrate how the performance incentives worked to reward the contractor for exceptional performance or to penalize the contractor for poor performance.

At this point you may be wondering what attributes and incentives have to do with FAST modeling. The answer is that they form a very important link to directed innovation through the FAST model, which will surface as we unfold the concept of linking attributes to contractual performance incentives. The study team decided that the

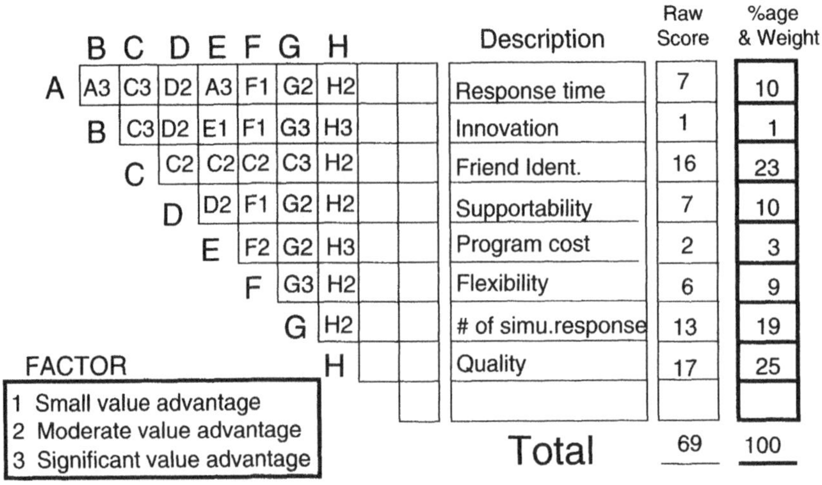

Paired comparison matrix:

	B	C	D	E	F	G	H
A	A3	C3	D2	A3	F1	G2	H2
B		C3	D2	E1	F1	G3	H3
C			C2	C2	C2	C3	H2
D				D2	F1	G2	H2
E					F2	G2	H3
F						G3	H2
G							H2
H							

FACTOR
1 Small value advantage
2 Moderate value advantage
3 Significant value advantage

Description	Raw Score	%age & Weight
Response time	7	10
Innovation	1	1
Friend Ident.	16	23
Supportability	7	10
Program cost	2	3
Flexibility	6	9
# of simu.response	13	19
Quality	17	25
Total	69	100

NOTE: *Innovation's scores was raised by "1" to keep the attribute in the system for evaluation. The other attribute scores were also raised by 1 to maintain their relative scores.*

FIGURE 6.27 Software Acquisition Process: Paired Comparison

Performance Incentive Using Attributes
Scale of Goodness

	ATTRIBUTES	WEIGHT	1	2	3	4	5	6	7	8	9	10
			Penalty Range			Target Range				Incentive Range		
A	Response time	10										
B	Innovation	1										
C	Friendly Ident.	23										
D	Supportability	10										
E	Program cost	3										
F	Flexibility	9										
G	# of simultaneous response	19										
H	Quality (MTBF hrs.)	25										
	Incentive Earned Points (IEP)											

FIGURE 6.28 Software Acquisition Process: Incentive Range

incentive range should start at 750 *incentive earned points*, with the incentive reward increasing in proportion to the score in that range. This is an adaptation of the scoring system discussed in connection with Figure 6.21. The higher an incentive earned points score of 750 and beyond, the higher the profit incentive reward. The target range of 400 to 749 is the expected or target range, so does not earn a performance incentive; in addition, a contract profit agreement is offered in that range. The penalty range includes incentive earned point scores from 399 to 100, with penalties proportional to the score. Reducing the contract's percentage profit in proportion to the performance shortfall enforces penalties. Additionally, the contract agreement would be in default if any of the attributes scored less than 1 on the attribute scale, regardless of the incentive earned points score.

Although it is mathematically possible for a contractor to score over 1000 incentive earned points, the chances of that happening are highly unlikely. However, in the very rare instance that seven attributes score in the 10 column, with, say, the eighth scoring over 10, the incentive limit would not exceed that of achieving an incentive earned points score of 1000 points.

Determining the Incentive Earned Points Score

The incentive earned points score is determined by multiplying the attribute weight by the column number indicating the performance achievement of that attribute. Figure 6.29 shows three cases and how their incentive earned points were calculated. In case 1, all the performance attributes fell under the goodness scale's column 1. Therefore, all the attribute weights were multiplied by 1, and the sum, 100, was recorded as the incentive earned points score for that case (e.g., attribute A would be 10×1, and attribute C would be 23×1). Case 2 is the same, except that falling under column 5, the attribute weights were multiplied by 5, for a incentive earned points score of 500 (e.g., attribute A would be 10×5, and attribute C would be 23×5). Case 3 is determined in the same manner as cases 1 and 2 except that a multiplier of 10 is used; the incentive earned points score totaled 1000 points. It should now be obvious that case 2, in column 5, yields an aggregated score that is five times larger than case 1. Similarly, case 3, in column 10, yields an aggregated score that is 10 times larger than case 1 and twice as large as case 2.

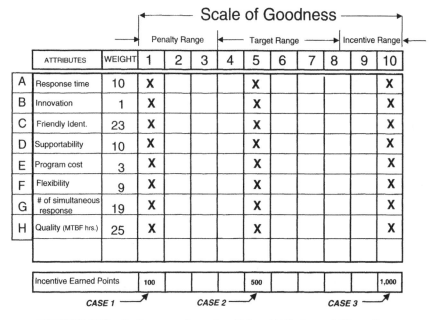

FIGURE 6.29 Performance Incentive Using Attributes and Three Cases

Case 4, shown as Figure 6.30, describes a probable result of a contractor's incentive position based on the performance of his product's attributes. This case shows a poor product performance, with an incentive earned points score of 257. With that score the contractor would be subject to a penalty in proportion to his score, which would reduce his expected profit. A quick analysis of Figure 6.30 shows that four (B, D, E, and F) of the eight attributes performed in the satisfactory-to-good performance range. Unfortunately, the good-performing attributes had low relative weights as determined by the customer's value preferences.

Are you beginning to see how a FAST model could help achieve higher incentive earned points? No? Let's continue. How can a contractor who is developing a new product channel innovation to focus on the more heavily weighted attributes? Prior to beginning the product development process and armed with weighted attributes that clearly identify the customer's perceived value, the contractor can create a FAST model describing the functions to be performed by the product. In creating the FAST model the product development team would identify those random functions that relate to the attributes as issues of concern. The FAST model developed by expanding and linking the random functions would then be dimensioned in matrix form (see Figures 6.4 and 6.25). Such a display would identify those product functions influenced or, as the prime driver, affecting the way the attributes performed. The product development team would then concentrate on creating function solutions that produced highest performance of the heavily weighted attributes. The results could very well be the situation described in Figure 6.31. The attribute profile described in case 5 in Figure 6.31 shows a total product attribute incentive earned points score of 741. The result is a happy contractor and a happy customer.

Although we recognize that balancing attributes is not a zero-sum game, there are many occasions in product development when achieving a high performance of one attribute would result in the automatic detriment of another attribute. Earlier in this

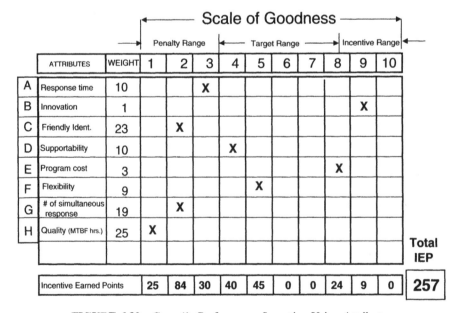

FIGURE 6.30   Case 4's Performance Incentive Using Attributes

Scale of Goodness

ATTRIBUTES	WEIGHT	1	2	3	4	5	6	7	8	9	10
		Penalty Range			Target Range				Incentive Range		
A Response time	10							X			
B Innovation	1				X						
C Friendly Ident.	23									X	
D Supportability	10					X					
E Program cost	3		X								
F Flexibility	9			X							
G # of simultaneous response	19								X		
H Quality (MTBF hrs.)	25									X	
Incentive Earned Points		0	6	27	4	50	0	70	152	432	0

Total IEP: 741

FIGURE 6.31 Case 5's Performance Incentive Using Attributes

case study we gave such an example, where an engineer's suggestion to trade attribute performance was rejected by the program manager. The reason for the rejection was the program manager's determination that *Friendly Identification* was valued less than any possible gain from improving *Response Time*. By assigning relative weights to the attributes through the paired comparison process (see Figure 6.27), the engineer would know if the proposed trade-off improved or lowered the product's incentive earned points score. For example, if we improve *Response Time* by one column, we gain 10 incentive earned points, but in doing so automatically degrade *Friendly Identification* by 23 incentive earned points, resulting in a net loss. (The really creative teams would try to find ways of improving *Response Time* using methods that would have no detrimental effect on *Friendly Identification*, as we discuss in Chapter 7.)

CLOSING REMARKS

Consumer and commercial product developers and producers should include in their market analysis the effort to identify market and customer highly valued attributes and link the need for improvement to their relative weight. This information is a tangible gauge which points to that elusive competitor indicator, perceived customer value. Armed with this information and a proficiency in constructing and analyzing FAST models would give the manufacturer another high-powered competitive weapon in its arsenal to gain competitive edge.

Using attributes as a FAST model dimension has broader applications than the case study example. Manufacturers offer "new and improved" product models for one primary reason: Their competitors have introduced something new that attracts customers which negatively affects others' sales and profits. The loss of sales is the result of one or more attributes losing perceived customer value. By determining which of a competitor's product attributes is the better performer or what new attributes changed the customer's value perception, a manufacturer can react quickly to isolate those attributes, and through analysis of the product's FAST model, focus on improving its performance, or adding new, more valued attributes in model upgrades.

The effects of responding to attributes in the consumer world are basically the same as using attribute performance as contractual incentive programs. Poorly performing attributes will lose profits through low sales. Conversely, the incentive in offering better-performing attributes is higher sales, market share, and earnings.

In Chapter 7 we explore how FAST models can enhance a team's innovative capability. Chapter 7 builds on this one, as we now have a means to define what customers want in a way that allows FAST models to be designed to achieve or exceed customer expectations.

CHAPTER 7

ENABLING INNOVATION

A variety of dimensioning techniques were discussed and illustrated in Chapter 5 and the reader was encouraged to experiment with other dimensioning schemes, being mindful of the rules that guide the drawing and reading of FAST models. In Chapter 6 we discussed how customer expectations of greater value could be made explicit and how this influences the scope and focus of FAST models. In previous chapters we have explained how to build FAST models and how to dimension them. We also linked the FAST model to strategic opportunity and problem framing by showing how issues of concern could be made explicit. So far we have shown how to make the FAST model an instrument for planning and managing projects as well as a stimulator for innovative thinking. Now we explore how to analyze the FAST model and how to gain insights that make creativity accessible and meaningful while creating value.

ANALYZING FAST MODELS

The care taken to ensure that the FAST model represents reality adequately in the form of a functional explanation makes dimensioning feasible. The functions, as verb–noun representations, should be abstract enough to enable alternative solutions to be considered but clear enough to describe functionally what should be done or how things are done. Remembering that, all we can ask of a function in an existing system is that it performs well. The FAST model offers an overview of the system under study, and dimensioning helps to reveal where dissatisfaction exists. It's not the functions themselves that we dimension and judge but the way a system was designed to perform those functions. When looking at the functions and their dimensions, performance metrics are selected to match the attributes chosen in the pre-event phase of the value study.

Stimulating Innovation in Products and Services: With Function Analysis and Mapping,
by J. Jerry Kaufman and Roy Woodhead
Copyright © 2006 John Wiley & Sons, Inc.

The performance metrics are how we assess the effectiveness of a solution that performs a function. Those metrics must be relevant to strategic value and the long-term viability of the organization rather than being focused singularly on short-term tactical or operational issues. For us, the notion of strategic value looks far wider than the internal considerations of a project. The project is thus a means to some larger and thus strategic ambition.

The metrics need to be objective and provide reliable information to observable and measurable solutions that combine to enable a system, or machine, to work. The FAST model is not created in isolation. It is an instrument born of a project that seeks to deliver organizational value; the project's goals and key attributes must be articulated as valued strategic outcomes. By relating to the functions in the FAST model the metrics chosen to describe the goals and attributes, the team maintains its focus on the expected outcome. It is by looking at the story being told by the performance metrics and the ideas they generate of why dissatisfaction exists that we are allowed to form a diagnosis and identify how things need to be done.

By spending time looking and discussing what the dimensions are telling us about the functional health of the things we have modeled as a complete system, we can gain insights into what kinds of innovations are required. If, as an example, some functions are seen as causing problems and declining sales, we must understand the conditions and the environment surrounding the problem. To illustrate, a company that produced thermocouples for the process industry was experiencing a steady decline in sales and market share. The project manager called for a design review and invested resources to upgrade the performance of the class of thermostats affected. Sales continued to decline even with a better-performing product. The project manager then convened a value engineering study team with the specific assignment of reducing the manufacturing cost. Since the objective was clear, the pre-event phase was omitted. The project manager's strategy was to reduce the price of the thermocouples in proportion to the cost reduction, thereby protecting profit margins and, through lower price, recovering lost market share. The VE cost reduction study was a success, exceeding its 25 percent cost reduction target with a unit cost reduction of over 30 percent.

Following implementation, the VE study resulted in a small sales recovery, but the general trend continued downward. The strategic needs had not been articulated adequately, so tactical ambitions were thus misaligned with long-term profitability. It wasn't until a new VE study was launched which included the pre-event phase omitted in the previous study that the real issue surfaced. This time the VE study focused on the root cause, the reason that sales were declining, instead of jumping to a cost reduction solution. As a result of the second VE study, the strategic problem was resolved and sales began to increase.

Following an exploration of those issues in a pre-event phase to make the causal relationships to sales more visible, a FAST model was constructed from order entry to delivery that reflected the entire sales process. The metric "time" was discovered to be valued higher by customers than "price." The time from order entry to delivery took about two weeks. Competitors were delivering replacement thermocouples to their customers in about three days or less. When a process shuts down because of a failed thermocouple, the cost to replace the faulty thermocouple is trivial compared to the revenue loss to the customer because of a halt in production. To improve response time to the customer, the VE team designed a thermocouple kit consisting of manufactured parts that could be assembled with probes of varying length. The design of the kits considered the different needs of customers serviced by each of the six regional sales offices. In addition to sales personnel, the sales offices were staffed with service people who were capable of assembling a needed thermocouple from the kit parts provided. This significantly improved aftermarket sales and service. In many cases a faulty thermocouple was replaced on the same day that the order was placed.

To improve the way that replacement thermocouples were ordered, service personnel on their scheduled visits placed adhesive labels next to their customers' units with the replacement part number and telephone number of the service representative. Labels were also placed adjacent to competing products, where such parts could be matched to make an alternative better known to the busy customers. With a greatly improved aftermarket order response time, combined with lower-priced products that resulted from the first study, sales accelerated, exceeding the previous sales volume record.

In another VE study we examined a production facility heading toward insolvency, only to discover that the salesstaff, hired on a commission-only basis, were encouraged to shift volume without considering scheduling implications and profitability issues. That is, the effectiveness of the sales team was assessed by the volume of product they placed orders for or by whether or not such orders were profitable to the firm. It turned out that a lot of the volume was for small companies and that the salespeople were giving large batches of free samples but that this was hidden among all the other transactions that were taking place. In so doing, the salesstaff exceeded their targets, and since small companies would never really need to place large future orders, management had inadvertently set up a systemic loop that kept everyone busy in non-wealth-creating activities. By unraveling what was supposed to happen, and what was happening, everyone gained a systemic view that showed how their individual efforts combined to form a system of dependencies that makes a firm work as a single entity. Until that shared organizational team perspective existed, the company was crippled by a myriad of theories of why things were going wrong. Labor productivity had regularly been blamed, with the result that union problems also emerged. The many disconnected notions of what was causing problems crippled management. The FAST model in this study allowed people in the various roles (e.g., sales team, quality control, production scheduling, warehouse management) to understand how they affect and are affected by each other's activities. Once an understanding of the concept of function dependencies had been achieved, many ideas were generated, often with no investment cost, which allowed new management responses to be designed.

If we were to reveal the ideas that helped the company, you would be forgiven for thinking that many of the suggestions were trivial. Often, ideas discussed outside the conversations in a value study seem facile, but to those involved they are often far from obvious, as they are blinded by the complexity of issues facing them. The company, which was facing a serious business decline, was hampered by quick-fix solutions. Those solutions did nothing to address the root cause, which continued to fester until those thinking rationally said, "Let's step back, take a breath, and approach this problem logically and systematically." Just as a doctor would recommend treatment only following a diagnostic investigation, the same is true for all reliable innovation processes. We argue that an examination of functionality is a feasible path to regular innovation.

Distinguishing Outcomes and Ideas

Where average life spans were once low, they are now extended because we use simple technologies such as soap, and more complex technologies such as medication. People are trying to adapt natural functioning so that our existences are more comfortable and enjoyable. Practical ideas about how to change things for the better and functioning are at their core. Just as we prefer good health to poor health, we prefer comfort to discomfort.

When someone asks for ideas, they are really asking for solutions, for ways to change the way that functions are performed. The functions can be from nature, such as the way that viruses spread and cause illness. They can also be from human-made systems such as having to prepare tax returns to finance government. The ultimate goal is to achieve a comfort level with which we are all satisfied, and it is that ambition that steers our notions of value and the recognition of innovation.

It is value that enables us to prefer outcome A to outcome B. So when looking at functions and asking for ideas, a response along the lines of "Let's clean up the environment!" is actually not a practical idea. It is a goal, an outcome that we would like to achieve because it moves us closer to the ultimate goal of a level of comfort with which we are all satisfied. However, just as a destination is not the journey, an outcome reveals little of how to achieve it. If we asked, "What do we have to do to get to that outcome?" so as to tease out the *how to*, we are searching for practical ideas. How can we be sure that the ideas will get us to our goal or that we have not missed an important step or broken a communication channel if we don't have an explicit and auditable functional representation such as a FAST model? We cannot see how innovation is possible, either objectively or intuitively, without linking an idea to some notion of functioning and functional dependencies or working that has to be done in order to change things.

All innovations are perceived as being better than what went before because they move us closer to the ultimate goals of human existence: to enjoy our lives and contribute benefits to the lives of others. Furthermore, before any improvement is possible, all innovations require a clear understanding of how things work. Therefore, an idea is a potential way of performing functions that leads to a desired outcome.

Starting to Generate Ideas

Rather than trying to generate a list of random ideas, by developing a FAST model we enable a team to appreciate the systemic need for innovation that is optimal for the entire system. It is because of this more complete understanding of how something should work, which is "learned" by the team responsible for delivering new ideas, that innovation can be realized systematically. To overcome professional and disciplinary boundaries and to bring the expectations of customers to bear, ideas are brought out into the open in a think tank type of environment. This allows organizations to manage ideas that may otherwise remain dormant or be lost within the complexity of the collective knowledge available to an organization if only it could isolate it and manage it adequately.

Having established a rich understanding, questions such as "How else can we perform this function?" or "How can we avoid the undesirable effects of this function?" enable a creative tension that is grounded in a systemic perspective. For example, to increase revenue a firm must sometimes lose business as it switches markets and stops producing old product lines. The loyalty that exists around old product lines often makes such switching difficult, as past success blinds employees to the fact that customers want new definitions of success. This is particularly the case with radical innovation, which requires a very different skill set to the one built around old product technologies with which most of the employees are familiar and comfortable.

The FAST model thus becomes a coordination mechanism in which we seek to find the best solution available by way of combining alternative ways of performing functions that reside within a system or product. A FAST model can help in making the need to switch core technologies obvious to employees. However, we must always be cautious, as side effects (e.g., lowering staff morale) may not be fully anticipated and so may diminish value. We must see our theories of which solutions to apply to functions as risk-based decisions. This now opens the door for decision technologies such as decision analysis[1] to be used to consider uncertainty as we set about devising experiments to check what we think will happen.

Just as a FAST model is a reflection of our understanding of how things work, so a decision tree is a reflection of what we are sure of, what we are not so sure of, and in the case of variables not in the decision tree, missed considerations due to ignorance. All these methodologies are intended to help us to manage the quality of our thinking. By making our theories explicit, we can test them. We must be sure to accept the fact that reality is far richer, more complex, and interconnected than we can ever represent with models. The FAST model can thus become a coordination tool for change. That it encompasses several disciplines allows us to progress beyond a single invention phase to consider multiple inventions from engineering to advertising within one innovation campaign and thus allows us to see FAST modeling as a more comprehensive innovation enabler.

This relationship between using functions to anchor the choices of alternative solutions allows a combining of processes that opens the door to *failure modes and effects analysis* (FMEA). It also brings FMEA into the conceptual design stages far earlier than is currently the norm. For example, if we design an engine for an aircraft, we may be able to recognize that proposed solutions to several functions could have catastrophic effects if they failed and so opt for an alternative solution that would reduce the risk. This consideration is easier to resolve while in the concept stage. Doing so would cut many months of research design and development (RD&D) time by more efficient modes of managing the way that teams set about thinking in the act of RD&D.

In our FAST models we often recognize causal side effects that we would like to avoid, as they detract from our desired outcome. In the next section we look at the concept of necessary but unwanted functions as *negative functions* and show how these can also be used to fuel innovation.

HANDLING NEGATIVE FUNCTIONS

When we discussed dimensioning in Chapter 5, we showed how it was possible to evaluate whether we were satisfied with a solution's performance of a function. This logic allows us to be able to consider swapping one method for another as coordinated by the function. Sometimes we also arrive at a situation where one function is necessary to counter the effects of solutions to the performance of other functions. For example, in cars a catalytic converter is a necessary solution to diminish pollutants from burning gasoline in an internal combustion engine. We all want efficient cars, yet we don't want the effects of pollution. So we try to control the undesirable effects of one solution by installing another solution to perform repairing functions. *Negative function* is a term we use to describe an unwanted function or the undesirable effect that results from a real-world solution used to deliver a core technology selected to achieve a basic function.

As stated previously, functions are not dimensioned. *Dimensioning* relates to the assessment of a solution selected to make a particular function work. Therefore, functions are never really negative or positive; those terms relate more to the performance of a solution. As such, *negative function* is a concept that implies a response to a solution or design that has been selected but that we would rather not have to deal with. Negative functions are therefore the consequence of design and the selection of technological solutions. They are not the major functions being addressed and so can never appear on the major logic path. However, they can be used to stimulate ideas that lead to innovations such as a hybrid engine, which reduces our dependency on gasoline as an engine fuel.

Examples of Negative Functions

The basic function of a projector, whether an overhead projector or a computer projector, is to *Project Images*. To achieve that objective, the light source in both projectors is an incandescent light bulb. The lumens produced to achieve the basic functions also produce the negative or unwanted phenomenon of heat. To overcome what we call a *negative function*, an additional device is incorporated in the design to *Dissipate Heat*. As such, *Dissipate Heat* would be a negative function. The solution's cost to perform *Dissipate Heat,* with its many collateral effects, is incorporated into the design not in support of the Basic Function but to counter the effects of the unwanted phenomena. We would simply like to ignore the negative effect, but the laws of nature and the worlds of physics, chemistry, and biology prevent us from doing so. Because the solution selected to *Dissipate Heat* is often expensive and noisy, the negative function is highlighted on the projector FAST model to call attention to it as a candidate for innovative investigation (see Figure 7.1).

Negative functions are always placed off the major logic path in the *when* direction. This is because the negative function is a consequence of the design rather than being selected intentionally to enable functions in the *how–why* of the major logic path to be achieved. As an example, when considering the energy efficiency of houses, we might select *Draught Exclusion* and then face the following functional conflicts: How do we *Condition* [home] *Environment* by *Regulate Temperature* and *Seal Leaks* to make regulating temperature an energy-efficient process? *When we Seal Leaks*, we *Trap Gases*. Trapping gases because we've have inhibited ventilation leads to an unwanted negative function. *Trap Gases* is the consequence of sealing and insulating efficiently a home to conserve the energy needed to heat and cool the home. To prevent the accumulation of potentially dangerous gases or smelly odors, the house must be ventilated. Ventilating a home reduces the efficiency of performing the function *Condition Environment*. What would probably happen is that we would seek an optimal balance between the heat output and heat losses for the purpose is concerned with ambient temperature and comfort. A practical solution to the negative function *Trap Gases* would cause us to open windows, thus undermining the money spent on draft exclusion; functions operate systemically. Technology advances in building materials and membranes are exploring ways of allowing accumulated gases to vent out of the home while preventing outside weather from entering. So we are witnessing functional

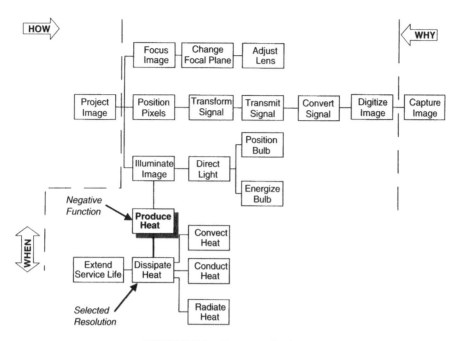

FIGURE 7.1 Computer Projector

innovation even if those involved are unaware of what they are doing, and doing so in ways that could be made more effective by creating a FAST model.

In some instances negative functions are so lacking in value that the main design has to be discarded and a different key technology searched for. The function *Extract Oil* in upper Alaska has an implied negative *Disturb Environment* function. *When* the environment is disturbed, you *Endanger Wildlife*. This function has prevented drilling for oil in the 19 million square acres of the Anwar region of Alaska until the negative function can be resolved. A potential solution may reside in drilling technology. With the advent of directional drilling, a large number of wells can be drilled from small platforms, each having multiple drills, occupying a small footprint. In this example it would require 2000 acres in the Anwar region to extract the oil reserves. This reduces the disturbed area significantly, to approximately 0.00015 percent of the region, consisting of a barren ice desert; the risk of pollution is now contained in a very small area.

The FAST model described in Figure 7.1 highlights the existence of a negative function. It reads: "When you *Illuminate Image*, you also *Produce Heat*." How, then, are we to handle that negative function? One way selected, as shown in Figure 7.1, is to *Dissipate Heat*. The FAST model now reads as follows: When you *Illuminate Image*, you *Produce Heat*. When heat is produced, you then *Dissipate Heat*. Heat is transferred by convection, conduction, and radiation. This basic physics can be used to get value out of a necessary but undesirable negative function. To address how best to dissipate heat, we drop to a lower level of abstraction by expanding that function and adding information that can help to generate innovative ideas, as shown in Figure 7.2.

A fourth option available for the creative handling of the *Produce Heat* function is to utilize the heat source in some useful way, converting the negative function to a desired *positive function*.[2] If through practical ingenuity we can find a use for a

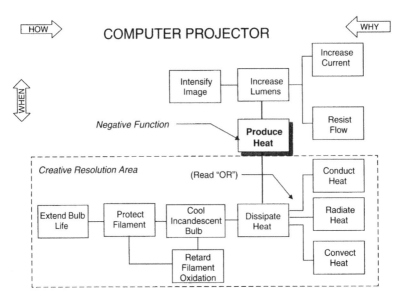

FIGURE 7.2 Lower Level of Abstraction

process that we must have, we give a "function" to that process and in so doing enable more value. This innovative ambition opens the way to combine FAST models with the *theory of inventive problem solving* (TRIZ). We now explore this integration of methods further to show how such a combination enables teams to access ideas from outside a workshop forum.

TRIZ and Negative Functions: Path to Creativity

TRIZ, a popular way to explore negative functions creatively, and classifies such functions as *contradictions*. Two examples of contradictions follow.

- Increase the diameter of the fan blades on a fan jet engine and reduce the noise generated.

We will look at a few examples of how TRIZ's view of contradiction can be used to convert non-value-adding phenomena and nonessential processes into valuable functionality. Normally, increasing the diameter of the fan blades, operating at the same rotational speed, will increase the noise generated; therefore, the two objectives are in conflict, or *contradict*.

- Increase the structural characteristics of a wheel on a high-speed rotating shaft while reducing the weight of the wheel.

The traditional approach to increasing the structural characteristics of a rotating component is to increase its thickness, which will result in increased weight; therefore, the above is also a contradiction.

Ophthalmologists searched many years for ways to repair detached retinas without using invasive surgery. This objective was considered a contraction. It wasn't until the advent of laser technology that the laser's usefulness, applied in a unique way, made this contradiction a solvable possibility. The laser wasn't developed for repairing detached retinas, so an inventive step was necessary. Putting the two together by redefining the problem in functional terms involves a high level of creativity, which we often call *practical ingenuity.*

It wasn't too many years ago that the removal of cataracts involved making an incision around the cornea to expose and remove the affected lens surgically. Postoperative care required immobilizing the head and eye for weeks until the incision healed. The contradiction was: How can we remove a cataract without intrusive surgery? The current procedure is to insert a small hollow needle into the cataract lens. High-frequency sound emulsifies the hard lens to a liquid, which is siphoned through the hollow needle. The technique reduces trauma to the eye significantly, so that the procedure is now performed as outpatient surgery. This led to performing an inner ocular lens (IOL) implant, to replace the affected lens, as a common part of the cataract surgery.

Essentially, this use of the term *contradictions* relates to outcomes that are seemingly in opposition to each other: for example, eat as much as you like (outcome 1) with a diet that causes you to lose weight (outcome 2); outcomes 1 and 2 seem to contradict their merger. As stated previously, outcomes don't explain the practical ingenuity needed, and this is where TRIZ uses software to search for ideas in external databases. TRIZ allows the investigator to "think outside the box" by searching for examples where the contradiction was resolved in a different industry. Let us look at some more contractions that are the same as our view of combining wanted and unwanted functions, which we've called negative functions and conversely, positive functions. A leading proponent of TRIZ, Ideation International,[3] uses a number of software programs to find and produce examples where contradictions, or negative functions, have been resolved creatively in patents.

Some of the software tools developed by Ideation International are:

- *Innovation situation questionnaire*: a tool for preliminary problem analysis that helps structure and document information about a problem situation into a format useful for problem solving.
- *Problem formulator*: a patented analytical tool that gives users the ability to model systems or problem situations in terms of cause-and-effect relationships and to attack the overall problem more effectively.
- *System of operations*: a knowledge base containing over 2 million patents worldwide. These include innovation "secrets" abstracted from successful results of previous inventions spanning a broad cross section of technological areas.

If we are faced with a contradiction as shown in the examples above, we can use tools from TRIZ as a resource to augment the search for ways to increase value from negative functions. The innovation team will study ways of adapting a solution from other, nonrelated industries, to convert a negative effect into one that increases value. This will increase our ability to innovate.

DEFINING PROBLEMS: PREREQUISITE TO SEEKING SOLUTIONS

Most solution seekers tend to form a preconceived solution based on the way a problem has been described. All too often, the way a problem is described contains a solution. Problem statements such as "We must reduce cost," "The time to develop new products takes too long," "Inventory turns need to be accelerated" are examples of solution directions embedded in problem statements. Such bias disconnects our need to explore a systemic perspective, so we may end up treating symptoms or even solving irrelevant problems.

Problem Set Matrix

Issues needing resolution consist of two elements: the problem and the solution. Determining whether the problem and solution are known or unknown can be described in a problem set matrix (see Figure 7.3). Each of the four quadrants in the problem set matrix reflects unique conditions intended to aid our discussion. This is intended to help distinguish situations, but in real life the lines between them sometimes blur and the reader must recognize this. When someone says "This is a set 3 type of problem," we must recognize that such is a management theory and explore why it is seen as being credible, so as to develop a means of testing whether the way we recognize situations is as good as is assumed. Remember that our underlying theory is that all models are intended to help us to think clearly and not to disconnect the need to inquire systemically. Understanding the conditions in the problem set matrix will help the problem solver pursue feasible solutions.

- *Set 1: The single right answer is sought.* The problem is known and a means to determine a solution is known.

This problem set describes a problem that can be solved with an analytical solution. The solution to this problem set has one specific value or "one right answer." As an

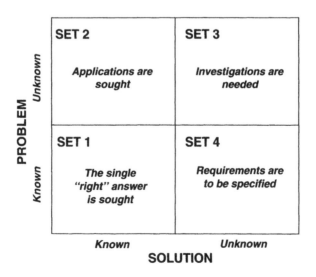

FIGURE 7.3   Problem Set Matrix

example, to determine the rate of gas consumption of an automobile requires calculating the amount consumed over a known distance traveled. The single answer is specific and can be determined by applying a known equation. Another simple example is to solve the problem, "What is 2 multiplied by 2?" The answer, 4, is a single specific value.

- *Set 2*: *Applications are sought.* The problem is unknown, but the solution is known.

This is commonly referred to as a solution looking for a problem. This problem set is common to management consultants, as their "shrink-wrapped" services are offered as a package and so represent a solution seeking problem opportunities and clients. Vendors waiting in the procurement lobby are another example. The vendors represent solutions looking for an application. Another example is the business of venture capitalists. They have the financial resources to assist emerging businesses to overcome their startup problems but don't know which startup company to invest in. Blue-sky research from our universities also falls into this category. The output of blue-sky research often results in solutions seeking practical applications.

- *Set 3*: *Investigations are needed.* Both the problem and solution are unknown.

A problem in this quadrant cannot be solved because it is not understood or perhaps even appreciated. Typical problem statements for this quadrant are "How can we meet the demands of the market?", "How can we improve the quality of our children's education over the next 10 years?", and "How can we avoid the need for intrusive surgery?" The problem placed in this set requires more fact finding and function definition to scope the problem and make it better known. This quadrant requires redefining the problem in a way that would identify paths leading to a solution. This is a highly creative task, but with a redefined problem, the investigation can then move into the set 4 quadrant.

- *Set 4*: *Requirements are to be specified.* The problem is known and the solution is unknown.

This problem set requires an investigation that will define the problem in clear and objective functional terms. To define problems in functional terms requires a high level of creative insight and enables the search for solutions to be expanded beyond the initial way the problem is characterized and subsequently viewed. By defining the problem in objective functional terms, problems and solutions that appear to have nothing in common can come together, and this realization precedes the search for an elegant solution. The situation described above, of matching laser technology to the problem of correcting detached retinas, is one such example. Once a potential problem-solution match has been made, requirements and specifications will tailor the search for solutions to the specific problem.

The awareness of negative functions thus allows us to be more sensitive to the possibility of further innovation, and that, after all, what this book is about. Let us continue on that theme and show how the FAST model can be used to target key functions to innovate.

Identifying Critical Innovation Points

The key advantage FAST modeling brings is that it highlights critical innovation points that if improved lead to significant and often breakthrough innovation. This means that we can see where to innovate with respect to the amount of change we can cope with. For example, if we choose to perform the basic function with a completely different technology in mind, every function to the right may become unnecessary. We touched on this earlier in the book when we talked about elegant solutions and the NASA pen that could write in space versus the Russian pencil.

Because the functions are coordinated by way of our logical intentionality, in the horizontal means–ends logic of the *how–why,* it is often possible either to seek major innovation by getting as close as possible to the left-hand scope line and to the basic function, or to seek minor innovations at the periphery of the FAST model. We have learned that once a function is designated as basic, it will never change. The way in which the system performs to achieve the basic function can change. Because of the dependency relationship between functions, changing functions on the major logic path close to the basic function will result in more radical system changes than will changing functions farther away from the basic function. Understanding this enables "change management" strategies to assess the capability for radical innovation or whether it might be better to develop an incremental approach to innovation and implementation.

As an example, if we determine that *Create Heat* is the basic function of the gas-filled cigarette lighter described in Chapter 4 (see Figures 4.18 and 4.19), the answer to the *how* question, *Produce Flame*, describes the method selected, in function terms, to perform the basic function. Note that a match performs almost all of the major logic functions. This shouldn't be surprising, since the cigarette lighter is a way to perform mechanically the functions of a match. Suppose that we wanted to radically change the way we perform the basic function. Is producing a flame the only way to create enough heat to ignite tobacco? The cigarette lighter in an automobile ignites tobacco by placing a resistance coil in an electrical circuit to create heat. Another approach would be to magnify and direct the sun's rays to the tip of the cigarette to create and direct heat. Replacing the *Produce Flame* function with either of the two alternative approaches will cause all the functions to the right of the basic function to change. The reason that all the major logic path functions change is because all the functions on the right of *Produce Flame* are logically dependent on that function for their existence. Removing and replacing that function removes those dependent functions selected to support the *Produce Flame* function.

REALIZING INNOVATION THROUGH FAST MODELS

Innovation is always perceived as an improvement on what went before. The way things used to be done is replaced with new approaches enabled by other inventions and new technologies in the larger act of innovation. When we map out how things are done with a FAST model, we abstract real-world phenomena, consider alternative methods and techniques in a conceptual world (i.e., in our minds) and then plan to implement our designs to create a new order and a new way of doing things in the real world.

Before the invention of the steam engine, horses limited the search for land-based transportation innovations. The airplane offered faster and better ways to travel across the oceans. From the beginning of human history we see the Stone Age giving way to the Bronze Age, Iron Age, Machine Age, Computer Age, and now the Cyborg Age. The history of humans is the history of how we find better ways to perform functions. This has not always been successful and our history is littered with failures that we can learn from. Ours is the story of practical know-how as we strive to reassemble raw materials and natural processes so as to enable us to achieve technological progress.

FAST models are about how to manage practical ingenuity that helps society and its organizations, projects, teams, and people to innovate and realize *progress*. Before we move on to the next chapter, let us take stock of what has been unveiled in the book so far by way of a case study. In the next section we present a background to the lessons we want to convey and then we present an actual case study for a product that did not live up to expectations in its operational use.

Toward Innovation That Makes a Difference

After reading this book and applying it in practice, many of you will experience the personal satisfaction of being in an innovative task team when their FAST model enables the breakthrough that leads to project success. Conversely, there will be times when other task teams question why so much time was used to build a FAST model. We live in an age of instant gratification and many people will undervalue the time and clarity of thought needed to codify knowledge and deal systematically with uncertainty.

Most FAST models represent the product or process being addressed by a task team. The FAST model is then analyzed to determine which functions to select for a subsequent *speculation phase* to find new and better solutions and achieve project goals. If we only build a FAST model without linking it to issues of concern and a strategic problem-framing episode, we run the risk of failing to identify the root cause and opportunities that are the prime reason for innovation in the first place. The purpose of a FAST model is not only to raise and normalize the level of understanding of how things work, but equally important, to focus on that part of the system that does *not* work or is not contributing enough value. Failure to focus on issues of concern compromises the purpose and value of FAST modeling as an effective contribution to problem solving.

To ensure that a project's issues of concern are being addressed in a project FAST model, information surfaced during the problem-framing phase translates into key issues that are linked to underlying functions. Using these functions to create a function model will result in building a relevant FAST model that will focus on those variables that if changed will yield a worthy difference or outcome. Focusing on issues also reduces the complexity and therefore the time required to build a specific FAST model.

IMPORTANCE OF THE PRE-EVENT PHASE

Picture the following situation. You are commissioned by the Home Bread Machines (HBM) Company to determine and resolve why HBM has been loosing money over the last three years with their new super breadmaker. If there were no systemic inquiry linking the product to the marketplace, it would not be surprising to assume that the fault lies in the cost of the breadmaker. The facilitator will then lead the task team in constructing a FAST model that describes the way the breadmaking machine works, functionally identifying and justifying the components of the product. The task team then uses the function–cost metrics displayed in the dimensioned FAST model to brainstorm ways of reducing the product's cost. Let's assume, however, that the design of the breadmaker is elegantly simple, exceeding approval body requirements and customers' expectations. Let's also assume that all members of the study team are infinitely familiar with the design, operations, and cost of the breadmaker.

HBM's business problems could be rooted in the pricing philosophy, manufacturing process, packaging, the organization structure, the amount of in-process and finished goods inventory, inventory turns, throughput, and a myriad of other conditions. A faulty distribution system and sales plan could also be contributing factors. If the project FAST model describes the way the product works, it will hardly contribute to resolving systemic problems. To be an effective contributor, the project FAST models must reflect the root cause and its effects as grounded in a larger and more interconnected system.

This example would also explain why some clients think that FAST is an excellent process for bridging problem identification with innovative solutions, yet others, who neglect to define the project properly in the pre-event phase, believe that the time spent developing the FAST model was a waste of effort. A good answer to the wrong problem has little value.

XYZ-3 CASE STUDY

The project selected as the case study for this topic is a communication device used by the military to display, analyze, and transmit field conditions. The project was selected for a VE study by the client because of the high cost of the product and its associated operation and maintenance cost. It is a portable electronics package easily carried, inserted, and removed from a variety of host vehicles. The main components are a structural case containing the system electronics, a removable hard drive, and a backup battery. Also included in the product are a display, keyboard, and interface cables. For the purpose of this discussion, the case study has been edited to focus on the topic and to protect the sensitivity of the project. The product, called the XYZ-3, has been in the field for approximately four years.

The XYZ-3 is a high-tech, state-of-the-art, complex electronic system. A FAST model identifying its systems, subsystems, and major components would be equally complex even if portrayed in a relatively middle to high level of abstraction. After completing the FAST model, it, like the previous breadmaking example, may fail to address the root problem or opportunity. Failure of the FAST model to describe and contribute adequately to the resolution of project issues often leaves the team questioning the value of mapping functions and causes the VE practitioner to lose confidence in the process. It is one reason that some VE practitioners are hesitant to use FAST models in value study assignments.

Defining XYZ-3's Problems

During the pre-event or problem-framing phase of the VM project, the answers to the three questions that characterized the problem are described as follows:

1. *What is the problem or opportunity that we are here to resolve?* The current XYZ-3 system results in acquisition and sustainment costs in excess of projected annual acquisition and ownership allocations.

2. *Why do you consider this a problem or opportunity?* Field failures have exceeded normal operational costs. Some common incidents include:

- Water intrusion into the power unit is reducing the product's operational life.
- Interface problems between components degrade reliability.
- Rough handling causes bezel button failures.
- Vibration causes electronic connectors to disengage.
- The system locks up and reboots too frequently.

3. *Why do you believe that a resolution is necessary? (or, What are the consequences of not resolving the issues?)* Failure to correct the deficiencies could lead to unacceptable field performance, thereby endangering the mission. Continued high cost will also affect the quantity of replacement field units. This could also affect the ability to meet the need for the XYZ-3 system.

It is important to see the link to performance, usage, and mission as part of a larger system within which the product functions. A non-systemic-minded facilitator might have seen the responses as a problem known–solution unknown type of situation and missed the point that not all the problems are in fact known or understood. In terms of the problem set matrix (see Figure 7.3) the team is facing a mixture of types. In the discussion leading to the answers to the three questions, much information surfaced as a source for identifying systemic issues and their underlying functions.

Three conditions became apparent in guiding the team to consensus with their answers to the three questions.

- Water, and probably other contaminants, are penetrating the system's case seals. This is a major cause of field failures.
- Field operators may not be following factory-prescribed operating and shutdown procedures, thus causing system damage and lost information.
- A larger than normal number of LRUs (line replacement units) pulled from field operations for repair show no sign of problems at the repair depot.

Setting Project Goals

The next step in the problem-framing process is to identify goals which if achieved would evidence resolution of the problem issues. The project goals of the XYZ-3 identified by project management were:

- Mean time between failures (MTBF) of field-replaceable units (FRUs) should exceed 2500 hours.
- Mean time between essential function failures (MTBEFFs) should exceed 700 hours.
- The rate of failures beyond fair wear and tear (BFWT) should be less than 50 percent.
- The water intrusion problem must be solved.
- Reduce the number of no evidence of failure found (NEOFF) units at the repair depot by 50 percent.
- Reduce the Turnaround time (TAT) of repaired LRUs.

The goals above flow out of the discussions and answers to the three questions and point toward concrete sources of evidence that could be used to assess whether subsequent improvement was being achieved. At this point in the process, it appeared that the way forward was either to revisit the product design, to investigate the handling of equipment in the field, or a combination of the two.

Selecting Attributes

Selecting, defining, prioritizing, and scaling key project attributes set the direction of the study and were used to define success in quantifiable terms (see Chapter 6 and the discussion of attributes). This exercise also established which attributes could be traded against other attributes if the team was faced with a conflicting decision necessary to improve value. The attributes selected, in descending order of priority, were:

- The number of built-in test (BIT) field failures
- The number of NEOFF units at the LRU level
- The number of failures BFWT
- The LRU cost of the processor unit
- The LRU cost of the system's display
- The repair cost at the operational and maintenance level
- The keyboard unit LRU cost
- The unit replacement cost of the removable hard disk drive cartridge

The process of selecting and analyzing project attributes led to the decision to divide the workshop participants into two teams. One team, representing the contractor, focused on the design characteristics of the XYZ-3. The second team, consisting of user or customer field personnel, addressed the operations and maintenance of the product.

Selecting Random Functions

Constructing a project FAST model begins with randomly selecting problem-related functions. The term *random* refers to selecting functions without regard to any preconceived dependency or sequence. (Function dependency is established during the act of developing the FAST model.) The functions selected are then expanded to add dependent (*how* direction) and independent (*why* direction) functions. When an adequate number (25 to 35) of expanded list functions have been identified, the FAST model-building process can begin.

Figure 7.4 illustrates the process of selecting and expanding random functions. Column 1 is a partial list of project issues of concern identified during the pre-event process. When the VE task team agreed that the major issues had been listed (typically, about 8 to 10 major issues), functions where articulated to best described what has to be done to yield an outcome that positively affects each issue (column 3). Then asking *how* and *why*, functions in the second and fourth columns were filled in. In the case study, both contractor and customer teams participated in developing the information in Figure 7.4, because the problem issues affected the design and operations of the system.

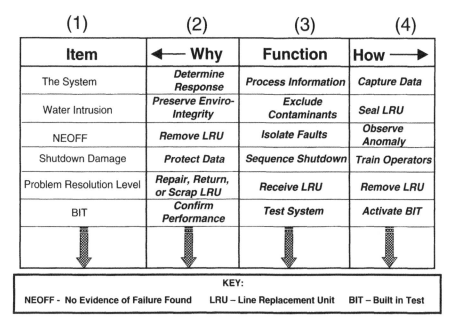

(1)	(2)	(3)	(4)
Item	**◄── Why**	**Function**	**How ──►**
The System	*Determine Response*	*Process Information*	*Capture Data*
Water Intrusion	*Preserve Enviro-Integrity*	*Exclude Contaminants*	*Seal LRU*
NEOFF	*Remove LRU*	*Isolate Faults*	*Observe Anomaly*
Shutdown Damage	*Protect Data*	*Sequence Shutdown*	*Train Operators*
Problem Resolution Level	*Repair, Return, or Scrap LRU*	*Receive LRU*	*Remove LRU*
BIT	*Confirm Performance*	*Test System*	*Activate BIT*

KEY:

NEOFF - No Evidence of Failure Found   LRU – Line Replacement Unit   BIT – Built in Test

FIGURE 7.4   Expanding Random Functions

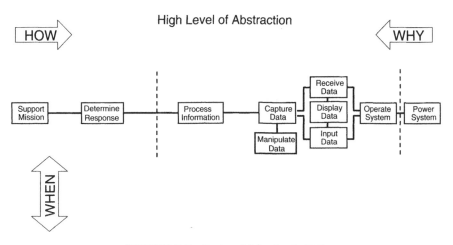

FIGURE 7.5   System Major Logic Path

Constructing the FAST Model

The functions in the expanded random function chart serve as a "starter kit" in creating the FAST model. The functions were copied to Post-It notes and placed on a large sheet. They need not be grouped in their originating sets of three, as shown in Figure 7.4. Figure 7.5 displays the major logic path created by expanding the functions from Figure 7.4. The spine, or major logic path, shown in Figure 7.5, functionally describes the intentionality for the XYZ-3 in a very high level of abstraction. The conditions we wanted to address resided in the activities and independent functions (*when* direction) that supported the functions on the major logic path. This is because the conditions selected for analysis are the consequence of the way the functions in the major logic path were to be implemented via particular solutions.

The individual Post-It notes with functions written on them are arranged on a large sheet of paper by the team as they follow the FAST modeling rules (see Figures 7.6 and 7.7). Additional functions are added. Also, some previously selected functions are edited to fit the *how–why* logic of the function map. Figures 7.6 and 7.7 show team members transposing the random functions onto a large sheet of paper as they begin to create the FAST model and check their logic.

FIGURE 7.6   Work in Progress

FIGURE 7.7   Building Shared Knowledge

When the FAST model's major logic path achieves team consensus, the team discusses the meaning, dependency, and location of the remaining functions. This process of asking and responding to the *how*, *why*, and *when* questions promotes the building and sharing of knowledge among team members. The task team selected the function *Ensure Operability* and placed it in the *when* direction to read: "When *Processing Information*, operations must be confident that the system is working to its performance specifications." Asking *How* resulted in an *and* branch. The response to "How do you *Ensure Operability*?" is answered: By *Confirm Performance* and *Preserve Environment Integrity*" (see Figure 7.8).

Figure 7.9 describes building off the function *Test System* shown in Figure 7.8. Asking "Why *Test System*?", the answer produced an *or* branch. Therefore, reading in the *why* direction, "Why do we need to *Test System*?" solicits "To *Confirm Performance* 'or' *Isolate Faults*." Continuing the logic questions produced the topmost branch of the FAST model. The completed FAST model, incorporating the issues of concern expressed in the random function exercise (Figure 7.4), is displayed in Figure 7.10.

As noted earlier, the symptoms of the problem expressed during the pre-event and strategic problem-framing discussions could stem from the system's design or the way that the XYZ-3 was operated and handled in the field. Therefore, the FAST models developed by the full participation of contractors and users in *co-development* expressed the concerns of both teams and were used by both to brainstorm potential corrective actions. Combining the knowledge of customers and manufacturers was possible because of the FAST model.

Selecting Functions to Be Brainstormed

Although both teams used the same FAST model, the functions selected for brainstorming and ideas developed varied, as they reflected the priorities and concerns of both subteams; remember that we split the task team into two groups. The shaded Post-It notes in Figure 7.11 show the functions selected for brainstorming by the contractor (i.e., product design team). Figure 7.12 illustrates the functions chosen by the field team (i.e., customer). It is important to note that when a function is selected for brainstorming, ideas generated should address not only that function but also the functions dependent on, or directly influenced by, the function selected. The circles drawn around the functions selected for brainstorming illustrate this.

Comparing the selection of functions to be brainstormed by both teams reveals that almost the entire lists of functions on the FAST model were considered for analysis. More important, those functions expressed the primary issues of concern to the project and enabled a sharing of practical knowledge that helped to realize a potential. It is from such conversations that we increase the ability to really innovate, and in a focused and targeted way.

Using FAST for Brainstorming

Each team member was provided with an 11- by 17-inch FAST drawing that they referred to constantly in expressing brainstormed ideas. This helped the participants to focus on the functions being addressed and how a particular function linked to other functions in the model. The shaded functions selected by the respective teams headed the brainstorming topics they then pursued.

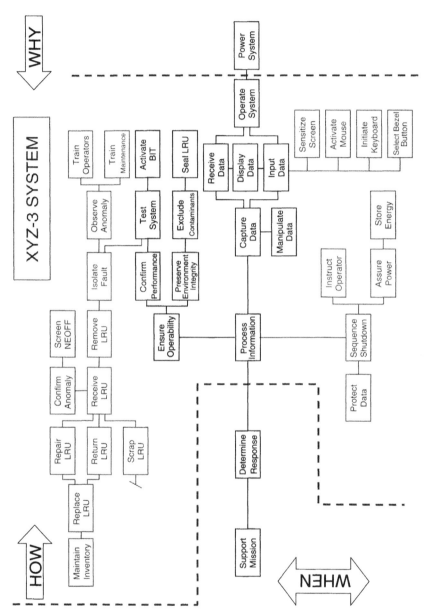

FIGURE 7.8 Building the FAST Model

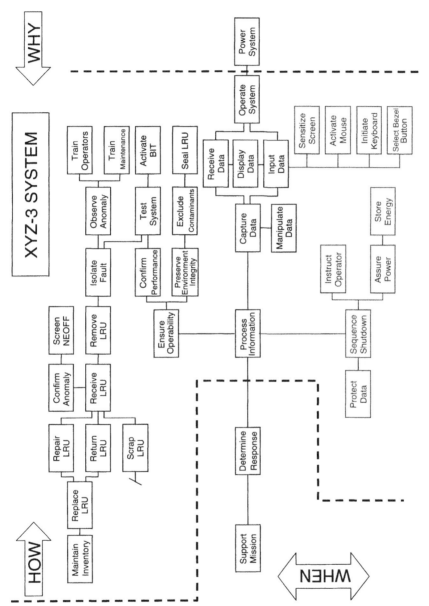

FIGURE 7.9 Continuing the FAST Model Process

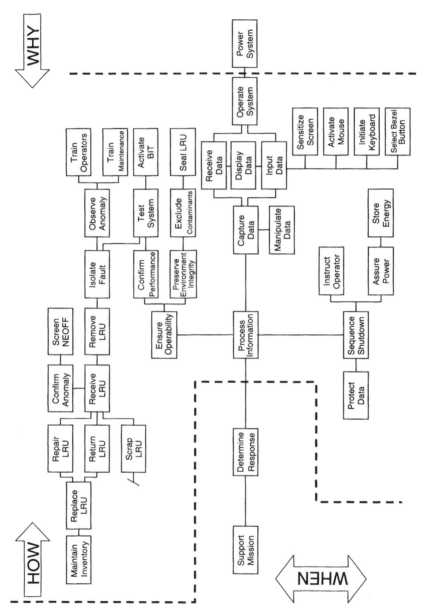

FIGURE 7.10 XYZ-3 System FAST Model

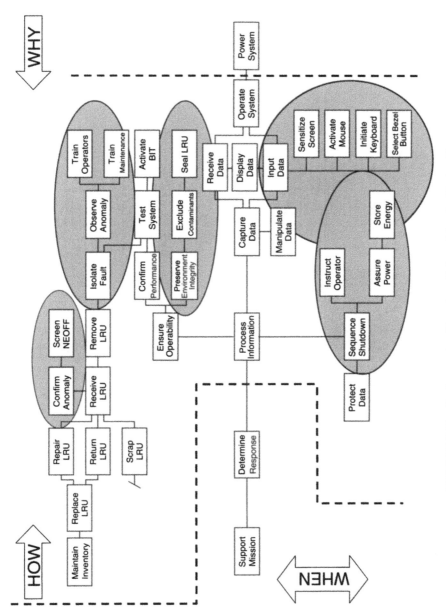

FIGURE 7.11 Brainstorming Functions Selected by the Design Team

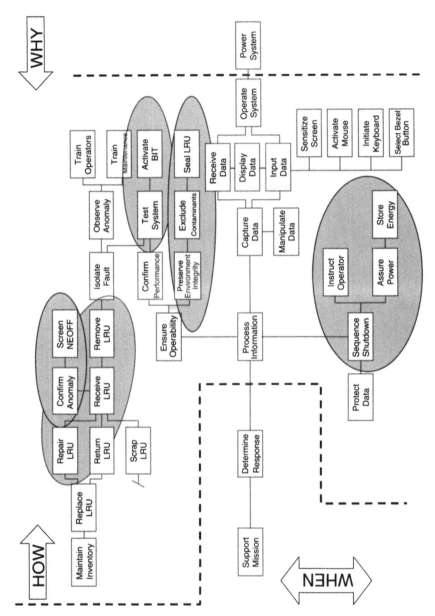

FIGURE 7.12 Brainstorming Functions Selected by the Field Customer Team

Concluding the XYZ-3 Value Study

After the ideas were posted, discussed, screened, combined, and summarized, the surviving ideas of both teams were entered on a common *proposal selection menu*. The consolidated team sought various ways of combining ideas and arrived at a number of proposal scenarios using the menu. Selecting the "best" scenario formed the basis for developing a business case to justify the proposed corrective actions. These proposals contained risk–reward assessed ideas that were combined to realize maximum value which subsequently were presented to senior managers for budgetary approval.

CLOSING REMARKS

In this chapter we have explored how the FAST model is used to direct creativity toward areas that if innovated would lead to a significant improvement in the overall system. This is possible because the FAST model is a tool that surfaces our team-based theories of how something works. As we learn how other disciplines explain their knowledge, our awareness of appropriate solutions becomes more wide. By addressing the parts of the system that need to be "improved" as the basis for value creation, the study team produces a FAST model that is a useful innovation tool rather than an interesting but ill-focused "wild idea" exercise. Therefore, selecting the focus for the FAST model and an appropriate level of abstraction, and articulating major issues and performance metrics, begin with the resolution of systemic problem framing in an innovation management exercise.

CHAPTER 8

FROM COMPETENCY TO CAPABILITY

In the preceding chapters we have unfolded a "how to" logic that develops a competency in mapping functions and building FAST models. As many people will attest, there is no way to shortcut practicing and experimenting on a variety of applications to develop proficiency. The more you work through the preceding chapters, the more your learning and competency will develop. After each new FAST model application, step back and examine the FAST model you participated in developing and ask, "What is this model telling me?" You may be surprised to find that principles are enforced and new insights surface that you can add to your FAST modeling capability. Furthermore, if you develop this capability as a team working on real projects, the shared learning will fuel discussions that heighten insights even further. In this chapter we look more closely at how we can derive greater value from the act of FAST modeling and so are less concerned with how to build them and more concerned with how their use unlocks greater potential for the entire enterprise or organization.

However, being a competent FAST modeler also requires other abilities, such as well-developed facilitation skills needed to guide a team through mapping and modeling functions. Rather than launch into a discussion about facilitation skills, we step back and consider why facilitating a group of people engaged in building a FAST model is necessary. We see the power of FAST modeling as an interdisciplinary team–based activity, so it needs someone to manage the technical and interpersonal communications. Rather than allowing information to pass linearly from one department to another, the FAST model enables a shared exchange of understanding and practical insights. We need facilitation skills to make the transfer of knowledge from experts into an explicit informational form, readily seen by others. To mold individuals with varying fields of expertise and interests into a workable team requires translating and combining different competencies into a new capability. Like individual players on a soccer team,

Stimulating Innovation in Products and Services: With Function Analysis and Mapping,
by J. Jerry Kaufman and Roy Woodhead
Copyright © 2006 John Wiley & Sons, Inc.

it is only by blending their individual competencies that the team is able to win or lose and so demonstrate what it, as a team, is capable of doing. The FAST model allows a collection of individual players to learn as a group and thus improves their ability to perform as an innovation team. The combining and blending of different types of expertise, knowledge, and experiences is a result we seek from multidisciplinary teams. But what purpose would surfacing and translating such knowledge serve?

It is not enough to create a high-powered team. There must be a purpose and reason that the team was formed. There would be little value if a well-trained, well-equipped, highly motivated soccer team entered the stadium ready to play, only to find that they are expected to play baseball. The test of FAST modeling competence lies in the insights that are unlocked and the innovation that follows. The successful FAST modeler is someone who makes a significant contribution to the success of an organization and its projects, products, and services.

MOVING TOWARD KNOW-HOW AND FAST MODELS

In this chapter we explore how to convert a competency in FAST modeling into a skill that leverages organizational capability. In this chapter we build on the knowledge and techniques of constructing FAST models and focus on how that act can be used to stimulate organizational wisdom. Being proficient in organizational wisdom and FAST modeling allows us to construct and manage supply chain models to meet the ever-increasing customer demands for newer, faster, better, lower-cost products and services. We are widening our view of functionality to include a collection of organizations that collaborate to bring products and offerings to the marketplace.

Let's recognize that organizations do not exist in isolation and that many of them trade with other organizations in supply chains. When a product arrives at a customer's home, many people working for different companies have performed functions such as *Assemble Product, Package Product, Distribute Product*, and so on. From a high level of abstraction we gain an overview of various companies working within a specific supply chain and performing dependent functions that span from raw material to consumer purchase. In some situations, through mergers and acquisitions, large organizations own and control how all the functions in such a supply chain are performed. In other situations, many small firms attend to a few functions and then pass on their solutions as contributions to an organization that assembles through the marketable product, work created by the functioning supply chain. In fact, when we look at companies such as Nike, the sportswear company, we realize that they don't actually manufacture anything and so control the product's value in a supply chain by controlling functions related to marketing and brand recognition. This is the level of abstraction that we deal with in this chapter as we discuss FAST modeling for organizational and competitive advantage. We begin by reviewing some key concepts in the field of knowledge management and then augment their views by bringing in philosophical considerations before discussing how FAST models can be used to improve projects and organizations as well as seeking greater value from supply chain synergies.

BEYOND INTUITION

Have you ever had some news and even though there is no way you could have known it, you feel as though you did know it? Perhaps you have had a problem and without understanding how you knew how to fix it, you somehow "guess" correctly. What about when you enter a room full of people who are not speaking and you "know" that they have been arguing. When someone throws a ball to you, do you make conscious calculations about trajectory, speed, and so on, or do you operate on automatic pilot and simply reach out and catch the ball? Have you ever thought about such embodied knowledge and wondered how and when you learned it? These are examples of what we could call *intuition*. We argue that the ability to describe what we know in words is less than what we actually know, and that this other knowledge is what intuition consists of. In the game of poker, expert players look for "tells," which are signals that another player sends out that can be used to gauge whether to bet more or to withdraw from the game. The point is that we perceive far more than verbal communication and that other senses are also sources of information. The various types of functioning can be at an individual level, such as in body language that is uncontrolled but conveys meaning. Or on a global level, some phenomena, such as a perceived threat that results from a nation developing nuclear technologies, may change the politics of a particular region. Intuition is part of a richer view of knowledge developed from a recognition of functions. The FAST model helps to surface the deeper intuitive knowledge as well as other types of knowledge.

There are many complex interrelating functions that make the world we live in what it is. Things that happen in the world of nature, such as atmospheric pressure and the weather, are part of nature's functioning. We become acclimatized to the world around us and so accept our environment and all the conditions that make up the environment. Although we can't change it, when we choose to study and observe such phenomena, our mind tries to understand how and why things happen in the world around us. We develop theories which evolve into principles that allow us to make use of these functions to build knowledge. That's how FAST works and the reason that the FAST logic questions *how*, *why*, and *when* (or *if–then*) are so profound.

When Isaac Newton made the connection between the weight of an apple and the fact that it fell toward Earth and in a straight line, he perceived this relationship in his mind by observing the world outside his own existence. The genius lies in recognizing the invisible laws through reason and experiment and then making them visible and accessible to others as laws of science. Such laws influence us as we build FAST models in the *when* or *if–then* direction, so we too draw upon previously gained knowledge when we attempt to model functions.

In his book *Knowledge and Its Limits*, Williamson[1] opens with the following paragraph:

> Knowledge and action are central relations between mind and world. In action, world is adapted to mind. In knowledge, mind is adapted to world. When world is maladapted to mind, there is a residue of desire. When mind is maladapted to world, there is a residue of belief. Desire aspires to action; belief aspires to knowledge. The point of desire is action; the point of belief is knowledge.

It was this paragraph that triggered the realization that FAST models had to deal with intention and desire on one hand (i.e., *how–why*) and on belief in scientific knowledge on the other (i.e., *when* or *if–then*). *Mind* refers to what we think as we see or sense it. *World* refers to reality as it exists and is therefore independent of our need, desire, or ability to observe it. Reality thus governs our models and not the other way around.

Reality is a complex interdependent functioning. The FAST model is simply a snapshot through a keyhole with a blurred lens. If we could see magnetism in the same way that we see a rainbow, the largest object in our sky would be Jupiter, not the Sun. That we cannot see Jupiter's magnetic field is functionally irrelevant to the decisions we make in our daily lives. The fact that Earth is protected from many meteor strikes as Jupiter's magnetic field sucks in those potential threats has no particular influence on our everyday decisions. Our survival is contingent on the functioning of enumerable things that we do not recognize. We see what our eyes allow us to see and build theories of reality based on our interpretation of what we observe. When we learn that it is possible to cause electrons to flow, as Faraday discovered, we become more aware of, and enlarge, our perception of the world. Our mind adapts its theories accordingly; we develop knowledge to advance technology. When a virus threatens us, we apply mind to the task of changing the condition of the world by building defenses to counter the virus or its effects. We are driven by the desire to form intentions, which lead to action. This is in the *how–why* logic of the FAST model. When things in the real world are not as we would like, we have a desire to change them. When the way things really are does not fit with our theories of how we see things, we question belief, as Galileo did when he argued that the Earth revolved around the Sun. When we build the FAST model, the need to corroborate the model with/to reality is important and is the basic premise on which the ability to swap one way of performing a function with a different solution is founded. The truth value of the FAST model is what makes it possible to achieve breakthrough innovations and significant value improvements.

We need to see at least two types of thinking to accommodate the world as is and our desire to change it to suit us better. The *how–why* of a FAST model is about our desires to change things. It's about intentionality. The *when* or *if–then* in a FAST model is how we represent our theories of how the world is "causally." Central to our book is the relationship between observation of what is going on and to people holding their observations as theories. This relates to what is really happening in the functioning reality.

When we delve into core theories from the field of knowledge management, we prepare the way for FAST models to become useful to senior managers and leaders. *Knowledge* is a word used so commonly that it lacks a consistent meaning across all disciplines. Some people think that it is the function of computers to store knowledge. But computers store data, which becomes information when sequenced to give meaning, which becomes knowledge when that meaning exists in the thought processes of people. The knowledge that water comprises two oxygen atoms and one hydrogen atom is a different kind of knowledge from knowing how much a cup of coffee costs in a local café. Both are examples of knowledge or they could be classed as information. Perhaps we should define and explain the differences among *data*, *information*, and *knowledge*.

In his book *The Haystack Syndrome*, Elyiehu Goldratt[2] defines *data* as "any string of characters that describe reality." He goes on to classify different levels of data:

- *Required data*: data needed by the decision process to derive the information needed
- *Invalid data*: data that are not needed to deduce the desired information

Information is defined eloquently by Goldratt as "an answer to a question asked." As an example, when having dinner at my favorite restaurant, I ask the waiter, "What do you recommend tonight?" If he answers "The roast lamb is very good tonight," that's information. However, if the waiter hands me the menu in answering my question so that I cannot easily make sense of what the best selection will be, that's data. To get information out of the menu I need to study the data presented in the menu, and that would lead me to ask more questions, such as "Is the salmon fresh?" The waiter's response could be in the form of additional data or information. When I am satisfied with the information, I can then make a decision. This raises the issue of the validity of information. *Erroneous information* is defined as a wrong answer to a question that has been asked.

In building a FAST model, we start by gathering random functions (see Figure 7.4). Random functions are data. The *how* and *why* questions we ask of the random functions (data) provide information. The power of FAST models is in building a shared body of information that allows a team to innovate, plot a course of action, and make decisions. Specific inventions and the resulting implementation that leads to innovation flow from a rich understanding of what needs to be done. It is no small feat that a FAST team is made up of a variety of professional disciplines, arriving at common agreements as to the informational content.

Knowledge is information stored that at some future time can be recalled to link with seemingly unrelated data to create new information. As managers of innovation, we believe that this is "only" what happens in the heads of people as they think. However, the philosopher Karl Popper argued that knowledge was objective.[3] Consider the functioning that takes place with DNA in the act of developing a human fetus. We see a similar view of objective knowledge as commands are seemingly enacted in accordance with an unfolding design. Such creation of new knowledge can be achieved with FAST models for high technology such as genetic engineering. Let us limit our scope here to general design: in particular, design that combines management with engineering and science.

The ability to innovate requires an ability to see how information can be used beyond the present application. So when a Scot named Sterling learned of Charles's law, which relates the volume of a gas to its temperature, he used the embedded principles as the basis of a curious engine known as a Sterling engine. This type of moving beyond "information" is often referred to as "thinking outside the box." People credited with being highly creative have this talent and are sought as valuable assets in building FAST models and in combining functions to create new ideas.

DISCOVERING NEW KNOWLEDGE

We are going to show how FAST models can be used to leverage innovation through know-how. It is our theory that innovation is the result of people sharing insights in social exchanges such as through conversations. Increasing the capability to innovate is a result of creating the conditions for fruitful dialogue. Innovation comes from knowledge within the thoughts of individual people. The FAST model is a communication tool for focusing conversations on practical ideas. Underlying this theory is a need to understand what is meant by the word *knowledge*.[4] It is only possible to "manage" something adequately if it is understood. In this section we look at what is meant by *knowledge*, which will help us to see the FAST model as a method for discovering new knowledge. A team builds the model, then through their conversations works out better solutions.

Drucker's[5] view that "you can't manage knowledge because it is between two ears" is how we see the relationship between people, their practical knowledge, and the FAST model. The FAST model is a representation of information created by structuring and sequencing data. The information becomes knowledge as people translate the information into their thinking processes, to innovate and effect change.

Knowledge and the discovery of new knowledge is what this book is all about. So let us define what we interpret so that you can glean the lessons and learn how to use FAST models to enable organizational improvement.

- *Implicit knowledge.* Ill-defined knowledge within people's heads.
- *Explicit knowledge.* Knowledge that we are conscious of and can speak of, such as knowledge of how to unlock a door. We classify information that influences the way we think and can be shared as *explicit*. For example, someone triggers a fire alarm and in so doing creates explicit information, which we hear, and our explicit knowledge encourages us to seek escape from a situation that we have been warned of but have not necessarily observed. In such a scenario, without our conscious awareness, our implicit knowledge would cause the release of adrenalin to aid our escape.

In the context of FAST models, we thus assemble people with implicit and explicit knowledge. Through the act of sharing explanations, we stimulate implicit knowledge to become explicit knowledge. We assemble the functions on paper in a logical structure that we extract from explicit knowledge in the heads of the team which we have encouraged be spoken out loud to build a learning cycle. Here we are talking about how better innovation intelligence is enabled by the team as it constructs a FAST model.

Nonaka[6] presents an explanation of how knowledge is transferred using an example of how a designer works with a baker to discover both implicit and explicit knowledge involved in breadmaking by a skilled baker:

1. First, she learns the tacit secrets of the baker at Osaka's International Hotel (socialization).
2. Next, she translates these secrets into explicit knowledge that she can communicate to her team members and others at Matsushita (articulation).
3. The team then standardizes this knowledge, putting it together into a manual or workbook and embodying it in a product (combination).
4. Finally, through the experience of creating a new product, Tanaka and her team members enrich their own tacit knowledge base (internalization). In particular, they come to understand in an extremely intuitive way that products like the home breadmaking machine can provide genuine quality. That is, the machine must make bread that is as good as that of a professional baker.

What Nonaka is talking about is how a craft-based solution to the functions needed to create bread can be understood and translated into a mechanical solution. The same functions that are necessary for the production of bread remain needed. What has changed is the way the functions are to be performed. The important point, though, is that some of the functions were not obvious even to the researcher or to the baker until the entire baking process was translated as functionally dependent links and those links were translated into the way the breadmaking machine was designed. The rigor involved in building a FAST model often forces these implicit skills to be defined as explicit functions.

The word *knowledge* as used here relates to the way that reality is manipulated by techniques. The innovation quest takes techniques performed by a human being and performs them through some kind of mechanical device. We see *function* as the underlying logic to create efficient and effective processes and systems. The need for any technique is to perform at least one function. That is why we see this book as contributing to a body of knowledge that seeks to increase the ability to innovate.

Dorothy Leonard [7] offers a view of knowledge when she talks of *T knowledge*. She explains that specialist knowledge is the *I stem*, in that it is deep, drills down, but is also narrow. It is the possession of particular and specific knowledge that we recognize as expert or specialist. To explain the difference between a generalist and a specialist, we use the letter T. The flat part of the T represents an overarching knowledge. This we refer to as *generalist*. There is an assumption that generalists are common and specialists are rare. It is because of such assumptions that generalists are often undervalued.[8] Both types of knowledge ownership, general and specialized, are equally necessary to enable a capability by their combination. The act of building a reliable organizational FAST model is a means of achieving a synthesis, or merger of specialist knowledge with generalist knowledge.

In Chapter 1 we talked about a group of chemical engineers whose shared expertise brought a focus that omitted consideration of other variables. We said that given a team all of whom were chemical engineers, there was a high probably that the solution recommended would be a chemical solution. The FAST model they would build would thus lean heavily toward the functions in chemical reactions and the necessary hardware and equipment. They may not have touched on sales or logistics, for example. Specialist knowledge alone is trapped inside what Leonard terms *I knowledge*. History is replete with examples of the folly of narrowly defining a situation when more consideration might have been wiser: from Marie Antoinette's "Let them eat cake" to Allison's multiple explanation of the Cuban missile crisis.[9] Conceding that specialist knowledge is important, it is generalist knowledge that creates the context in which specialist knowledge is applied *Context* is what gives knowledge value. Knowing how many angels can fit on the head of a pin seems worthless because we know of no context in which such knowledge would be useful. Remember our earlier view that *value* relates to usefulness that leads to a better state of affairs. This is why the problem-framing function of the pre-event in a VE study and the definition of what we actually mean by strategic value is so important for the FAST modeling process.

Specialist knowledge and expertise are viewed more favorably than generalist knowledge. In large companies, this leads to fragmented knowledge domains of specialist knowledge, centers of excellence, departments, and eventually, management silos.[10] The fragmentation of knowledge, combined with differences in salary levels, amenities, and power structures, affect the willingness of one department to reveal insights to an internal rival from another department. The resulting decisions therefore cannot be made with the best intelligence available. Words such as *team* or *collaboration* hide the fact that getting on in life is awarded to persons who excel. The emergence of *communities of practice*[11] (CoPs) is an attempt to join knowledge fragments together and break through the walls that separate a "line and staff" approach to knowledge location. Many CoPs will gravitate toward expertise and become intimidating places to ask naive questions. Unless generalists are members of CoPs, knowledge will remain trapped inside professional languages, values, and cultures.[12] The FAST model, with its clear line of sight to purpose and profitability, provides a means of combining specialist and generalist knowledge and thus makes better use of the social capital that exists within an organization.[13]

Experts are generally found in staff positions in large companies, headed by department managers. A program manager focusing on the business metrics of a major contract, or product line, often heads this line position. The staff department manager assigns temporary specialists, with a particular competency, to serve on a line program to support the objectives of the program manager. The program manager may have a number of staff members with different specialties on any given program. This is regularly called a *matrix organization*. However, its effectiveness does not flow from organizational charts but from the way that information and know-how is shared. This is why the FAST model has such a successful history in helping project teams to spot breakthrough innovations and ideas that are so obvious that no one can figure out why they were not considered and implemented before.

A common problem arises when the assigned specialists' staff manager's goals conflict with the goals of the project manager. The specialist's professional growth is in the hands of the staff manager, but the program manager judges the performance of the specialist. It is therefore important that although having separate interests, staff managers and program managers must have supportive and dependent goals that address the larger business and strategic issues of the company. A common strategic issue is to make money now and in the future. The FAST model is able to help teams to visualize and bring about such dependences between line and staff departments. It's about making sure that the functions of the organization are served by the best management solutions.

The dimensioned FAST model, undertaken with a collection of people with different types of knowledge, becomes a tool to aid team-based reasoning. The RASI (see Chapter 5) becomes more than a tool to identify who should be in charge to make sure that things don't go awry. RASI becomes a way to redesign teams so that the knowledge available is utilized in the best way. The FAST modeling becomes more than a value improvement technique. FAST modeling becomes a means to generate strategic insights that lead to organizational advantage.[14]

Organizational units are designed to perform functions. As such, they cannot exist and achieve *purpose* in isolation. When studying functions we learned that there is a dependency of functions, just as a mechanical component depends on the functions performed by other components. This concept of dependency in mechanical devices is equally valid in organizations. When Peter Drucker gave us management by objectives (MBO) in the mid-1960s, it was embraced by managers as a "silver bullet" solution to rising productivity. The thought was that each organization's management layers should be measured against performance goals. Although the approach seemed logical, the concept often failed, for two principal reasons. First, the hidden assumption in the MBO approach flowed out of the view that it was a top-down process: that is, senior management sharing their business goals and commitments with subordinates, allowing them to devise their own ways to structure their work to achieve those objectives. The intent of MBO was to align subordinate goals so that they support management's performance commitments. In practice, senior management, in effect, asked their subordinates: What are you going to do this year that will warrant your advancement? Once the subordinate goals were established, the goals, which translated to performance measurements, dictated the behavior of the workers.

Business is dynamic. What was an important objective at the beginning of a business fiscal year may have lost its priority to more urgent directional market changes by midyear. However, knowing that their career advancement depended on meeting their committed goals, the more urgent business issues received little attention, because accountability was not focused on addressing the new and often important issues. Second, although the subunits had the best of intentions in accepting challenging goals, they soon found that to achieve their committed objectives, they were very dependent on the performance and information of other subunits. A salesperson committed to annual sales volumes depends on production personnel to have products available to sell. The underlying functions of the organization were not performed adequately because a consolidated theory of functioning had not been articulated. In this book's wide sense of the word, rather than narrowly implying function as a management role, the fact that functions were never made explicit is the reason that the MBO project stalled; they lost sight of the organizational functionality.

If an organization's highly structured chains of command undermine the ability to achieve innovation, we need a dimensioned FAST model to expose barriers if we are to understand the sources of lost opportunity. If on the whole an organization needs clearly defined structures and processes in order to function with external organizations, it must also enable creativity and alternative modes of working or it will become unable to adapt to changing environments. What we are talking about here is far more than workshops and other kinds of "days away from the office," which no one really believes will make a difference. What we are presenting is a process to form an organization that would be designed to best serve the functions it was intended to perform. We are arguing for FAST modeling to be used by senior managers to enable organizational designs as a collaborative endeavor within the organization.

The clearly defined structures, roles, and responsibilities as seen in an organization chart or in quality assurance manuals reflect solutions to functions that an organization must perform. When procedures are designed outside an explicit view of functionality, the designer would be guided by a desire to ensure responsibility, accountability, authority, and a means of control. This type of thinking yields regulations and boundaries on which departments and departmental staffs often base their identities; this is the type of logic that ends in management "silos". The stereotypical joke, "Hi, I'm from the finance department and I'm here to help you," gives us a glimpse into an underlying truth we all know but don't necessarily bring out into the open and examine.

An organization needs to perform very different, often conflicting functions simultaneously. It needs to have clearly defined structures and to demonstrate a regulated capability; at the same time, it needs to have the opposite, so that new ideas can surface, be nurtured, develop, and grow (see our discussion of negative functions in Chapter 7). Rather than allowing internal politics and interdepartmental budgets to undermine capability, senior managers need a better way to think about the design and operations within the organization. They need to know what really has to happen in order to be successful. The FAST model allows different processes and procedures to be considered not in a detached and isolated way but with respect to the well-being of the internal organization and the business it serves.

MANAGEMENT OF FUNCTIONALITY

We have to have a theory regarding what is best to do and what is not. The challenge before senior managers is to understand the different types of knowledge at play and to combine them to best perform the functions of the organization. The FAST model enables a group of people to articulate the necessary practical and scientific knowledge to achieve organizational goals. Furthermore, as the organization itself exists in a functioning system called the "marketplace," it needs to direct its practical and scientific knowledge toward survival, success, and growth.

Organizations are a form of technology comprising internal and external functioning. The concept of an organization exists (functions) for at least a single purpose (outcome) or it would not endure. If senior managers can make that purpose explicit through the development of a FAST model, they are better positioned to function as great leaders. To many, the ability to make money for shareholders will be primary. Satisfying customers thus becomes a means to that end. But what is the obligation of a company to the community in which it resides? The profit motive is strong, but the organization is also an entity that creates employment and social structures as well as stability in local communities. Here, then, is another example of function dependency. As a company grows, so does the community. This necessitates expanding infrastructures, services, housing, schools, and other support structures. The community is dependent on the company for its sustainability, which in turn is dependent on the community for its resources. The scope lines of a FAST model must be attuned to the appropriate width of consideration that is an appropriate strategic framing of the context for innovation.

The organization performs functions in order to achieve a purpose or purposes. A number of perspectives exist around general purposes such as progress, wealth creation, consumer benefits, social value, and so on. We need to step back and recognize that each possible purpose is contextual and ensure that our FAST model reflects them. By *contextual* we mean the bounds we place around our gaze, such that if we look at, say, taxation, the context would be the interface between government, the public and private sectors, and individual taxpayers and their democratic right to vote a high-tax, low-value government out of power. The lens you are looking through makes a particular outcome seem like a plausible purpose. If you are an investor, making money for shareholders makes sense as a purpose. If you are a parent trying to earn wages to feed your family, social value to community makes sense. The same will apply to other possible purposes. The multiple views are caused by the different ways that people attach significance to what they perceive, and that leads to certain kinds of explanations. This multiple-perspective theory is what we mean when suggesting that people look at a single reality through different lenses. Here there are no multiple realities; there is only one reality perceived in multiple ways.

To manage is to make the complexity visible and understandable (see the case study for construction management students in Chapter 6, which makes a complex theory explicit). This is what a FAST model is capable of achieving by representing how an organization and a supply chain are converted from the implicit into the explicit. The question we need to ask of organizations being observed is: Where does the lens we look through originate? Here we recognize systems within systems. Leonard[15] describes a firm as comprising (1) a physical system (tools and methodologies), (2) managerial systems, (3) skills and knowledge, and (4) values.[16]

A FAST model in this context must reveal functions for all those systems. Many organization charts and the placement of organizational functions within those charts are dependent on the "comfort zone" of the chief operating officer. A senior manager who trusts and is comfortable with a subordinate manager may shift that subordinate manager and his or her staff so that they report directly to him or her. It isn't surprising that when the COO leaves, the first move of the new COO is to reorganize the organization to fit his or her comfort zone. The task, though, is not about the person but about whether the functions that need to be performed will be performed well and win approval from internal and external stakeholders. The FAST model could model personal functions related to the individual or choose to ignore them and concentrate on the essential functions needed to run the firm. Although some management theories reduce the emphasis on people, such as the scientific management adherents, others, such as Mary Parker Follet, promote a view of management that is more inclusive of human values. From this we must acknowledge that our views on the topic of management itself provide a guide to what we select as relevant. So we must test our biases to ensure that our FAST model really is aligned with an informed view of the issues of concern.

By seeing capabilities of organizations as the collective knowledge, desire, and commitments, temporary organizations such as task force teams can be designed to address and resolve multifunctional issues quickly and effectively. The way forward is to see the FAST model as a means to convert implicit knowledge (e.g., unarticulated know-how) into explicit information and knowledge and to use such a conversion process as a means to test and develop shared views of espoused and actual theories of action[17]: in other words, to use the process of developing a common theory of functioning as a way of surfacing, testing, and combining our theories of which functions need to be performed and how.

Using FAST Modeling to Improve the Supply Chain

Because an organization does not exist in a vacuum, its external relationships must also be considered. By getting a group of people who represent the key functions of an organization's supply chain to collaborate in building a FAST model from raw material, through end-customer purchase, to end-of-use disposal of the product or service, we can enable insights that are impossible from the viewpoint of individuals and individual companies within that supply chain. The initial FAST model need not convey any sensitive data. It simply demands that the valuable contribution provided by each node in the supply chain be made explicit and shown in terms of intentionality and dependency (note the *how–why* representation here) and causality (the vertical *if–then* or vertical *when* representation). This teasing out of implicit theories of the supply chain, in and of itself, will allow a strategic vantage point.[18]

Remember, a value-adding process is one that has a function. For mature firms in supply chains, if firm A performs function x (e.g., machining) and firm B performs function y (e.g., finishing), this may be more to do with the history of the firms and the supply chain's evolution than with an efficient design. Another supply chain that is in competition with firms A and B may have a firm C that has a cost advantage by performing both functions x and y in-house. In other words, it is only by considering which functions need to be performed that the level of value added can be assessed. This is what business process reengineering missed.[19]

When designing a new supply chain, attention should be focused on the functions that must be performed within the boundaries of that supply chain, without regard to who will perform those functions. Only after the supply chain, represented by the major logic path in a FAST model, achieves consensus of the interorganization team that created the model can the team discuss how the chain should be divided by supplier capabilities that can effectively perform those functions. The result is fewer suppliers in the chain, better supply chain control, faster processing time, better in-process quality, and a lower end-product cost. Supply chains should be assessed periodically to determine how new suppliers and emerging technology can improve the overall performance of existing chains.

If those involved in a supply chain can be made comfortable sharing data, the FAST model can be used to design a more elegant supply chain and the necessary processes within it. This willingness to share privileged information may mean that some firms will be designed out of existence in order to achieve greater effectiveness. This will flow out of the merging and elimination of redundant or overlapping processes in an evolutionary approach to industry innovation that gives way to a revolutionary approach founded on functionality and the pursuit of elegant solutions. The ability to develop and honor trust is a key competitive advantage, for only if a "win–win" principle is in place will inefficient firms enable other firms to optimize supply chain value. As managers from many companies collaborate to develop a function model, they are forced to expose their processes and procedures and expose them to analysis by others. Those dependent on other companies will also have the opportunity to define the "inputs" needed from the dependent companies to function more effectively. As this unfolds, it is common to hear statements such as "Is that why you guys do that? I always wondered about that" or "No that's not what really happens. It's what is supposed to happen, but we never actually do that because. ...." And so on. The conversations around FAST modeling allow the illusions we have to be tested against evidence.

Using FAST Modeling to Enable Shared Understanding

The following section is organized against the background of a case study for an exploration and production team from the oil industry challenged with improving the production rate for some oil wells. As you read the case study, think about the following management problem. Imagine that you are charged with improving the production rate of existing oil wells. You form a multidisciplinary team comprising investors, accountants, managers, health and safety staff, engineers, geologists, and others to achieve an innovation breakthrough. Each discipline has its own set of values and identities, culture, and technical vocabulary. How would you determine if the plan that emerged was credible? Mull this over and later (in Figure 8.7) we show how we tackled this problem in a project that needed to find more creative ways to increase oil production by fracturing rock at the bottom of the oil well.

Managing Intangible Value to Advantage

Let's begin by seeing the way that organizations create value from a task-oriented perspective. The sequential "do job x, then do job y, then do job z" view of management processes is often seen in the disciplines of project management. The project management lens prompts us to use such words as *deliverables* and sees its role in terms of hard, measurable outcomes. What is often assumed is that the purpose is known,

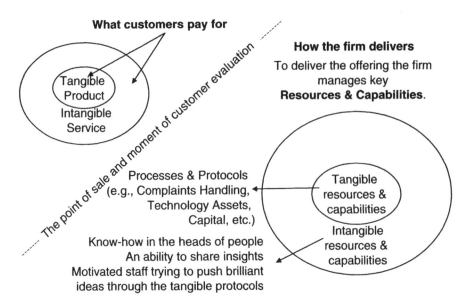

FIGURE 8.1 Knowledge and Value Creation

but within complex projects using many tiers of suppliers, the core purpose is often lost. Each element has its own way to measure incremental performance, which may conflict with the higher purpose of the project.

The organization seeks to earn money from customers in exchange for the firm's *offering* (see Figure 8.1), that is, a tangible product such as a new car or a new drug (see the top left-hand corner of the figure). Surrounding the tangible product are intangible services such as those provided by a shop assistant, or how key personnel help customers, or how project managers rush around to ensure the deliverables are on time, within budget, and within specs. To innovate here would be to improve the physical system with its current tools and methodologies and as a result modify management systems. However, customers often want far more than "on time, in budget, within specs," so the role for project management is limited to task-oriented thinking and projects are limited to the deliverables the customer expects.

Look into the organization and you will find different types of knowledge. At the heart of any organization are people, and as shown in Figure 8.1, it is their actions that determine how the firm delivers the offering (see the bottom right-hand side). People have varying levels of competence. It is the ability to synchronize these varying levels of competence to create organizational capability that is a major challenge in the new economy. Just as a group of the world's greatest musicians cannot spontaneously become an orchestra, there is a need for a generalist who enables the various specialists to "harmonize." Compared to an organization, managing an orchestra is much easier because feedback is gathered easily and quickly. Either they play well as an orchestra or they do not. But an organization is not so easy to assess. Sometimes the things that really count cannot be counted easily, so do not emerge as issues. Feedback becomes filtered and appears as symptoms and effects of larger problems. Solving the more visible symptoms allows the problem to fester and emerge in another form in another place in another time. This is why the act of building FAST models is so important.

So with employees, we have tangible resources leading to organizational capability. If a manager needs a secretary, we might assume that any secretary will do. However, as many executives will explain, not all secretaries are equally competent. Employers have knowledge in and around certain disciplines that is embodied within particular people. Furthermore, many people will have inklings and implicit knowledge of which they are not even aware. As we dimension the FAST model, such considerations surface and are essential. Around the tangible resources and capabilities are intangible resources and intangible capabilities (see the bottom right circles in Figure 8.1). Combining that knowledge with a self-starting ability (i.e., high levels of initiative), the secretary relieves the executive of many time-consuming administrative tasks. As more administrative responsibilities are assumed, the secretary's position will rise to match the capabilities of the person they assist, and the secretary will become an executive assistant. Many secretaries fill the expanded role of executive assistant, performing much of the data and information gathering needed by the manager to make decisions. The FAST model can reveal the underlying functions so that they are made explicit. Conversations between an executive and his or her assistant during FAST model construction become more meaningful and help to raise performance standards because of the heightened understanding that follows such clarity of communication.

The organization functions through the development and application of all kinds of knowledge—a far richer array than in the procedural views of knowledge espoused by business schools. The knowledge extends from the sincere salesperson with a greeting that really does make a customer feel valued, to the R&D scientist who realizes that an electric current causes a certain molecule to either attract or repel water molecules, to the engineer who recognizes a practical use for such scientific insight to create a moneymaking product sold by the sincere salesperson. The organization could have the smartest engineers and scientists in the world but if the salesperson cannot win the trust and confidence of customers, success will not be achieved. In a well-appointed restaurant run by the best chef, if the service staff is not trained and the waiter insults a customer, the restaurant's business life will be short, regardless of the food quality, presentation, and ambience. Organizational success comes from the collective efforts of all employees dedicated to success, which must be defined in such a way as to win the "buy-in" of all the people serving the company. The know-how in the heads of people is shared when commitment to organizational success is believed to be more relevant than the need for defensive withholding of essential information (see Figure 8.2).

The tangible and intangible resources and organizational capabilities make possible the tangible product and the intangible services that project management deals with. Modeling functions provide a way to synthesize the generalist and specialist as well as technical and nontechnical insights and different managerial roles and responsibilities so that an organization can begin to use *transformation* as a source of competitive advantage.

Customers decide the effectiveness of the entire offering (see Figure 8.3). They decide value as they can choose between different offers in the marketplace. Customers can express what they don't like by rejecting the sales department's offerings. But they often have difficulty articulating what they do like about the offer selected. It is for this reason that customers must be well represented in the development of FAST models, including their views. They must be represented directly either in person, through focus groups, or through customer research so that uninformed assumptions about

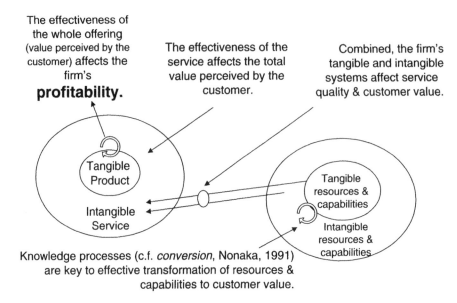

The effectiveness of the whole offering (value perceived by the customer) affects the firm's **profitability.**

The effectiveness of the service affects the total value perceived by the customer.

Combined, the firm's tangible and intangible systems affect service quality & customer value.

Tangible Product

Intangible Service

Tangible resources & capabilities

Intangible resources & capabilities

Knowledge processes (c.f. *conversion*, Nonaka, 1991) are key to effective transformation of resources & capabilities to customer value.

FIGURE 8.2 Systemic View of the Tasks and Deliverables

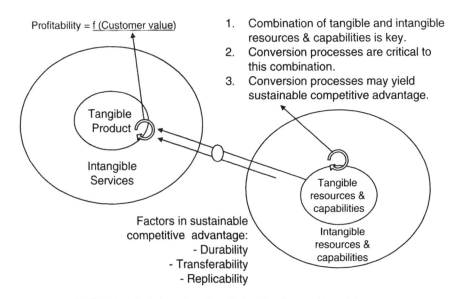

Profitability = f (Customer value)

1. Combination of tangible and intangible resources & capabilities is key.
2. Conversion processes are critical to this combination.
3. Conversion processes may yield sustainable competitive advantage.

Tangible Product

Intangible Services

Tangible resources & capabilities

Intangible resources & capabilities

Factors in sustainable competitive advantage:
- Durability
- Transferability
- Replicability

FIGURE 8.3 Managing Knowledge for Competitive Advantage

what customers value are minimized and replaced by more reliable information. We cannot assume to know what functions customers will value. Profitability thus becomes a key goal for commercial organizations and how their projects enable profitability. Project management looks after the effectiveness of the service that leads to the tangible product. This is augmented by an organization's service quality, aftermarket services, and a dedication to deliver customer value. It is the ability to unite all the various types of knowledge in profit-oriented conversations that allows us to move more deeply into what is possible.

Rivals could visit an organization with such advantages and be unable to see the source of competitive advantage born out of FAST modeling as a collective team. It's not how fast, how cheap, how accurate a rival is; it's how willing the employees on the production line are to prevent defects, to maintain throughput, to share ideas, to help another team; how good their understanding is of the role they play in a supply chain with shared destiny. Such can only result from shared understanding and common agreement to purpose. It's about how good employees perform key functions that should be in the organizational FAST model, and this capability flows from a sense of shared destiny built out of the conversations that take place as the FAST model is developed.

AUTOMOTIVE PARTS CASE STUDY

Figure 8.4 describes an automotive parts manufacturer's business system. The company designs and produces components for new automobile models as well as supplying parts for the aftermarket. The company, formally part of a larger company, was spun off to form its own independent venture. A basis was needed upon which to think about the entire business so that they could plan strategically. Following this FAST model, not only could they consider ways to improve internal processes but could also look at which functions could be outsourced and which must be kept embedded in their basic product offering to protect their competitive advantage.

The FAST model is presented in its raw form. That is, the model is not dimensioned. Missing from the model are the assigned responsibilities and accountabilities (e.g., RASI) and other dimensions that would make this FAST model case specific to the problems or opportunity issues that the particular senior management team wants resolved. We show it here to get you to think about the conversations and knowledge that result from it. Without the FAST modeling, such conversations would not happen with a team working together in the same room. The act of FAST modeling creates the need and purpose for focused conversations that cut across disciplinary and organizational boundaries.

In Chapters 4 and 5 we described how to construct and dimension FAST models. At this stage of the FAST model development, what dimension schemes can we use to make our FAST model more useful?

1. The functions can be clustered into departments and their boundaries for organizational realignment in which the functions of the departments reside. What issues do you imagine might be discussed?

2. Performance goals can be assigned to each department. The FAST model will display which departments depend on another department's output to achieve their performance goals. How could one department learn about another without such a tool that fosters dialogue rather than defensive communication and the need to defend turf?

3. The goals of each department can be assessed to determine their contribution to meeting the company's goals. How else could senior management ensure that departmental planning had common intentionality and was heading in the same direction?

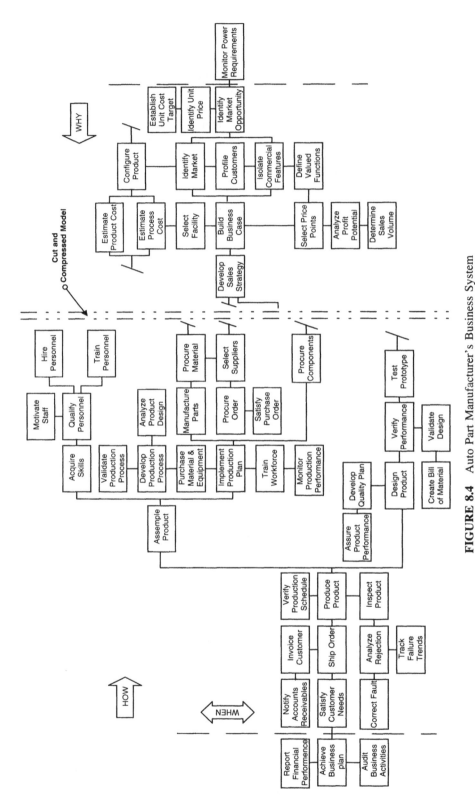

FIGURE 8.4 Auto Part Manufacturer's Business System

192

4. Key output performance dates can be placed on the model to identify any potential constraints. Here the role of the customer feeds into the discussions, as the team is focused on meeting the expectations of those customers and how decisions must track backward from key milestones.

5. A RASI matrix below the FAST model will make explicit areas of accountability, responsibility, support, and oversight by individual disciplines within departments. In what other ways could we develop a rich view of what has to be achieved by way of functionality and how the current organizational design is suited to those requirements?

6. Color-coding the functions as "performing well," "performing poorly," and "not performing" will provide an overall assessment of the company's performance. The results will indicate areas that need improvement. Isolating and expanding those areas in lower abstraction levels will help uncover the reason for subperformance. Such information placed on the FAST model can make it a useful tool for management teams, especially for progress chasing during implementation.

7. A study of the FAST model will help identify how to respond to major market changes and changes in the business environment by identifying those functions most affected. The model enables a dynamic view to be elicited and management scenarios to be generated in such a way that organizational intelligence is heightened. How many other techniques are available that enable a collection of people to build a shared understanding of what functions need to be performed, what solutions currently need improvement, how the people in the firm are involved with the entire organizational process, how and why individuals contribute to the system, and whether innovation should be radical or tame? This is why the act of FAST modeling, done correctly, is a powerful tool for senior management.

HOW CAN WE UNIFY?

The act of building a FAST model is where its core value rests. It has value in other uses, such as in progress chasing or in allowing competitive benchmarking. It is a stimulant of rich dialogue that cuts across organizational and departmental boundaries. The key problem is that all employees start off from different perspectives with respect to any shared vision statement crafted by senior management. This is made more complex because of professional vocabularies. To one group "stress relief" is jargon, and to another it has rich meaning. We need to get individuals to teach themselves as a team. The FAST model provides a way to model methods from different departments, disciplines, and companies into a single operating system of what is supposed to happen. Such a system can enable a clear line of sight from the contribution of a person or machine or process at the detailed operational level, to the profit or loss that will be felt at the executive level of the organization. By building a single philosophy in the form of a FAST model, the participants argue "No, that's not what happens; I'll tell you what happens" and in so doing move implicit knowledge through information and into explicit knowledge in a process that has an objective truth independent of any person. The quality of the model is a reflection of how easy or difficult it is for people to reject what is wrong or false. Building an environment where intimidation is suppressed and self-esteem protected is an environment where new ideas and practicality will reign.

Here we combine the methods we have discussed with a collection of persons who use collaborative inquiry and think as a team.[20] It is important to recognize that all statements are tentative and should be treated as such. They should be treated as quasi-theories or working hypotheses that need testing. Although we might believe them to be correct, we still search actively for corroborating evidence that is unambiguous. If someone says that *"abc* is a function of *xyz"* or "The reason we *Question Customers* is to *Secure Feedback*,"[21] we must see such as management ideas of what is happening, and why. Whenever someone makes a statement such as "What we really need to do is . . .", they are offering one way forward from many alternative ways. Implicit in the statement is an unspoken management idea of why this makes sense. It is common in organizations to ask *how*. The question *why*, which is rarely asked routinely outside the context of the FAST model, and provides the reason and justification for actions linked to functions. It is thus a key benefit of building a FAST model to surface the ideas we often don't articulate, and if we do this as a team, we are ready for double-loop learning. Asking *why* of an unwanted condition is also a key element to uncovering the root cause responsible for that condition. If we can expose some of the deeply held theories, question why they have credibility, and search for observable evidence, we can get a group of people to learn as a team as they generate ideas, test them, modify them as necessary, and figure out why they seemed plausible when first generated.

We have to recognize that what we see as significant is a product of our awareness and the priorities that we hold in our heads. Furthermore, these priorities are about moving the person closer to a better existence. Value is directed toward improvement. What *better existence* means is decided by individual people, such as a commuter looking for a quicker journey, a mother wanting more time with her children, a scientist searching for a cure. Priorities are about how best to perform functions with available resources, and usefulness is often a pragmatic determinant of which functions we focus attention on and which we allow to slip past us.

The act of building a FAST model is about unfolding our ideas and explaining why we believe them. At one end of a scale mapping out "how we shall believe" are gullible mindsets in which people believe everything they are told. Gullibility is about accepting both that which should be accepted and that which should be rejected. At the other end of the scale is extreme skepticism. Extreme skeptical mindsets reject everything. They lead the person to reject that which should be rejected but sadly also to reject that which should be accepted. So a skeptic will be a skeptic in all conversations. It is difficult to imagine anyone seeking to apply gullibility as a consistent framework.[22,23] Perhaps some people are skeptical for reasons that we cannot grasp.

An approach to countering skeptics and reducing the number of negative responses when soliciting ideas for ways to implement the functions in the FAST model is to rule that judging an idea cannot begin with the phrase "no, because. . .." This recognizes the common condition of rejecting the goodness of an idea for a small fault that may exist within it. A "no, because" response tends to reject the entire idea without recourse. By using the expression "yes, if . . ." when judging ideas, the fault is isolated as an issue that needs to be resolved before the idea can be considered. In Figure 8.5 we provide examples of the differences in judging ideas using both expressions.

FUNCTIONAL ENQUIRY

We begin our investigation into personal beliefs by asking people to be aware that no one has irrefutable proof that their way of seeing reality is correct or complete. We want

Negative Response	Positive Response
NO, BECAUSE .. • *The proposal is too risky* • *It's against company policy* • *Management won't buy it* • *It won't meet requirements* • *They'll never approve the investment* • *It will create a health hazard*	**YES, IF ....** • *The risk can be managed* • *The policy can be changed* • *A sponsor can be found* • *Requirements can be changed* • *We can make the returns on the investment attractive* • *The hazard can be controlled*

FIGURE 8.5   Judging Ideas

to create a genuine attempt to get those involved to build new solutions from shared insights as their FAST model takes shape. We see the universe as a functioning reality and that our consciousness exists in that functioning reality. The history of humans is about how they have manipulated the functioning reality to form the Stone Age, Iron Age, Bronze Age, and so on, through the Steam Age, Computer Age, and on to the Cyborg Age we are now entering.[24] Functions related to food, shelter, and mobility have always been central to humans. Other functions, such as *Product Demand*, are a result of a socially constructed modern society with its systems, subsystems, and so on.

Functioning is far richer than our FAST models can truly represent. The FAST model is at best a rough approximation of a rich complexity that functions in ways that we humans glimpse. When Henry Ford conceived of the assembly line, which led to the mass production of automobiles, a need for roads, car parks, service stations, and traffic lights emerged. The core function was about enabling mobility, and the Model T was a solution that fitted with the disposable income affordable in a mass market. This combination made other functions more significant. What was previously an acceptable road for horses now lacked the necessary requirements for cars. Humans rise above nature's limitations by inventing machines and social systems. However, these inventions cause other dependent functions to become necessary as an idea triggers an innovation process. As one key invention increases mobility, expands eating experiences, improves comfort and luxury, enhances communication, lengthens a high-quality life span, or brings exhilarating entertainment, many subsequent dependent functions become necessary. As we invent and then innovate, we take on the responsibility of functioning in the complex, interacting reality within which we exist. I once heard someone lament: "If I knew that the copy machine would become so popular, I would have invested in file cabinets." We now have the means and ability to create a paperless society, but paradigms have deep roots, making it difficult to break free and create another paradigm, to be tested again, with future technology advancements. This is the rich functioning background that FAST models attempt to reveal.

When we say such things as "If customer complaints keep rising, we are going to lose market share," we describe a causal relationship that tells us what is going on with a function,[25] or in the case of falling sales, not going on. We have to make these explicit in the form of FAST models to understand the link between intentions and causal relationships. It is this type of enquiry that allows us to identify issues of concern facing management.

Once we believe that we understand how a system creates value, our inner belief system will prioritize data. Our FAST model represents reality in the form of structured information. But we must accept that the lens through which we look may distort or bias our view of reality. To prevent potential errors, we seek corroboration from external evidence. The way we do this is to hold the FAST model against a system dynamic (SD) model (see Chapter 6 and, e.g., Figure 6.8) and simulate the effect that a particular change will have in the way that a function is performed. Clustering groups of functions in a FAST model easily represents this. Such clusters represent an existing system in a system dynamic model. It was information that emerged during the problem-framing session (i.e., pre-event) that guided our investigation of the issues of concern. An example of clustering functions is offered in Figure 8.6. This FAST model describes the high-level functions of a hospital from the point of patient admittance to that of satisfying the concerns of the patient. The model uses clustered functions as a way to distinguish between the three major activities performed by physicians, nurses, and administrators. The FAST model also shows the linking functions that unite the three major activities.

Displaying the clustered functions establishes the department boundaries and the functions within those boundaries. In our generic hospital operations FAST model, each department is managed differently. Also, they had very different performance metrics. However, if the major issue of concern is how better to *Serve Community* (the higher-order function), the FAST model provides a visual display allowing representatives of the various units to collaborate in determining new solutions to support the higher-order function. The FAST model also shows that the three departments cannot operate in complete isolation. There are dependent function links between departments that support and contribute to each department's performance.

The act of building a FAST model and the rich dialogue enable us to consider swapping solutions used to perform individual functions and run simulations to see whether they actually make a difference at the system level. This process of generating alternative methods to perform functions and then trial-and-error substitution is *single-loop learning*. The real advantage is to figure out why we believed that previous decisions were good or bad. That is, we use the act of innovation to force collaborative learning. This yields insights that can be quite startling, and in a supportive environment where people can talk openly, it can become a source of sustainable competitive advantage. This is because such inquiry examines organizational capability (i.e., how a firm's design is able to perform organizational functions efficiently and effectively). Sometimes we learn that high levels of professional competence are trapped inside "silos" and wasted.

The FAST model for an organization should span end of product life through customer selection at the point of sale and back to getting raw materials. By mapping the complete value stream and identifying the functions performed, the organization can see which parts of the supply chain need to be innovated, closed down, or owned. By exposing all the intentional functions performed by an industry network or cluster, we can identify which ones create value, which could be innovated, which should be outsourced, and which need to be produced in-house.

Here we use *if*–*then* rather than *when* to associate causality and constraints. Figure 8.7 shows a simplistic FAST model from the oil industry exploration and production challenge that we asked you to consider earlier. We asked how you would combine the various types of knowledge. Here we show how that was achieved

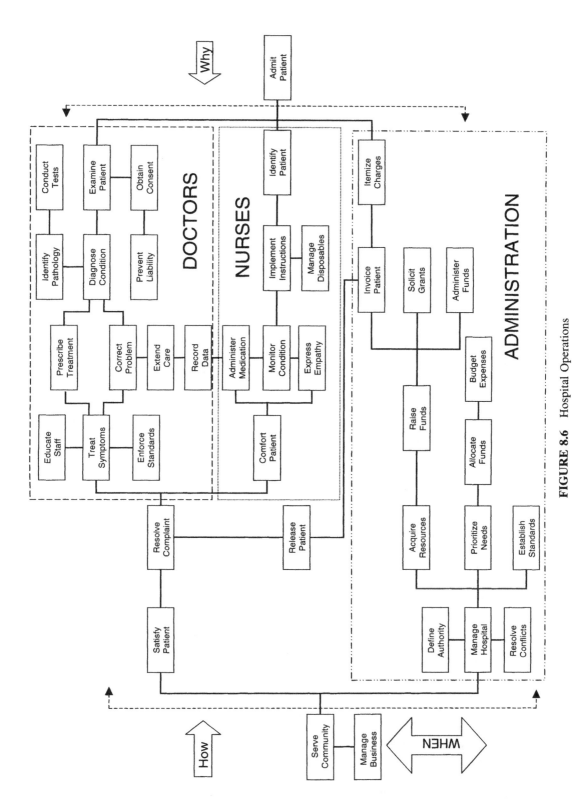

FIGURE 8.6 Hospital Operations

Articulate *in Teams*: What is done, why and how; widen knowledge.

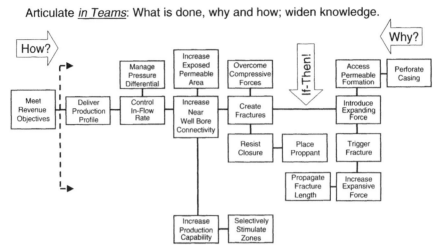

FIGURE 8.7   FAST Model to Coordinate Team Thinking

in a small project. It is important to note that different disciplines can grasp what has to happen and why, so their implicit and explicit knowledge combine after rich explanations in the workshop environment. The resulting decision process is more informed. The reason we use *if –then* in place of *when* is that our goal is not primarily to stimulate innovation as in previous examples. The goal here is to develop a more reliable representation that will allow senior managers from different firms in a supply chain to gain an insight that will encourage negotiations and alliances that enable greater effectiveness. The unidirectional logic of *if –then* increases the likelihood of the model's truthful representation of what actually happens.

Figure 8.8 shows another example of the *if –then* rule used to explain how a diesel engine works with compression ignition; there are no spark plugs in a diesel engine. The key difference between Figures 8.7 and 8.8 is the scope of consideration. The scope lines identify the boundaries of the study, which is determined by those building the FAST model and recognition of the fact that customers search for the basic function before considering other functions. (Note that there are no scope lines in Figure 8.8, which signifies that the FAST model would continue in either direction.) It is in the problem-framing sessions that issues of concern are raised which are so vital to the building of FAST models.

Referring back to Figure 8.4 and our discussion about FAST models for organizations and supply chains, if we dimension a reliable representation of the functions performed in a supply chain, we can design policies that benefit the collaboration between firms that are not obvious at the single-organization level. We can thus use FAST models to assess the various types of knowledge at play, the skills and competencies needed today and in the future, the resources available, and the capabilities the supply chain is able to muster. By gaining such a perspective the supply chain can negotiate with local universities to design educational programs, to develop interfirm technology benefits, to remove unnecessary insecurity costs that exist between firms not aware of their shared destiny, to build social connections that rivals in low-labor-cost countries will find difficult to replicate, and so on. The FAST model will enable a strategic vantage point that shows how various organizations contribute value to the

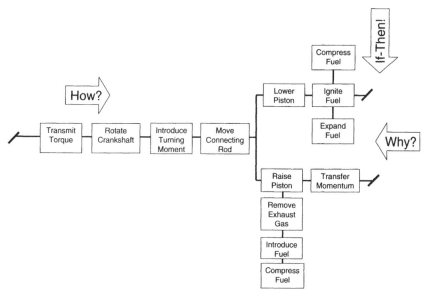

FIGURE 8.8   Use of *If – Then* to Condition Intentionality

supply chain and at a lower level of abstraction, how particular roles, such as quality control in a particular firm, provide a necessary benefit to the entire supply chain. It is not the size of a cog in a machine that determines its value, it is how such a component enables the machine to function. The FAST model enables us to make our implicit knowledge explicit and shared and in the context of a supply chain how an overload of often competing interests collaborate to steer behaviors and actions that seek profitability and the advancement of progress. Furthermore, the act of building a FAST model also forces us to identify areas and topics we know little about but with reflection realize that we should inquire into further.

In review, we have taken you through a "how to" series of chapters that reveal how you can become more competent at building FAST models so as to manage invention and innovation. In this chapter we have explained how, by seeing organizations as human-made systems that exist in networks or supply chains, it is possible to apply FAST to model functions for interfirm innovation and in so doing increase the capability of the organization and the supply chain.

It is by stepping back and making our management theories of how things work explicit that we tap into implicit knowledge: by seeing views of what should be done, or otherwise. Management needs a guiding framework that links how we think (i.e., mind) to reality (i.e., world). A function makes a necessary and valuable contribution to the system of which it is a part. If a thing has no function, it has no value and can be removed with no undesirable effect. This is what guides us to elegant solutions. To design processes and systems and products and services with no exploration of function is to start designing in an inefficient and ineffective way.

CLOSING REMARKS

We hope that this chapter has shown how FAST modeling can help you and your teams transcend single-loop learning and become double-loop learners, as a team. FAST will enhance your organization and its projects to achieve greater levels of value. Efficiency must always be considered with effectiveness in successful FAST studies. The reason for this is that concepts of purpose and function are linked in the act of design. It is this triad of purpose, system, and function that when combined through the act of building a FAST model becomes open to invention and innovation.

The ability to build an appropriate FAST model and facilitate the necessary discussions is the ability to help organizations to innovate systematically and achieve greater levels of value. We hope that this book helps you to unlock the potential that lies trapped inside organizations.

END NOTES

CHAPTER 1

1. As an aside, Larry Miles also met regularly with an industrial psychologist, Clare W. Graves, who later went on to describe the taxonomy of human coping systems we know today as spiral dynamics. It is worth noting that these two disciplines, value analysis and spiral dynamics, coevolved from lunchtime conversations between two men in Schenectady. What is also interesting is that one of them looks at how to innovate and the other at how various culture types respond to the notion of innovation; they look at different parts of the same innovation agenda.

2. For the story of Larry Miles and the emergence of value analysis and function analysis, see O'Brien (1987).

3. O'Brien (1987).

4. Here we cite some key references from the fields of value analysis, value engineering, and value management that give a sense of the development of the discussions around function analysis through FAST diagrams and on to dimensioned FAST models: Akiyama (1991), Bytheway (1968), Fallon (1980), Kaufman (1990, 1998), Miles (1961), Snodgrass and Kasi (1986), and Woodhead and McCuish (2002).

5. Here we bring in discourse from philosophers of science and philosophers of technology to show a deeper and richer series of discussions that we use to underpin the practitioner's theories and which allow us to move forward to new theories that enable systematic approaches to innovation: Cummins (1975), Kroes (1998), Mahner and Bunge (2001), Preston (1998), and Wright (1973).

6. Here are some conference papers that marked the arrival of a new approach and use for FAST models whose practitioners did not feel that they needed to be

Stimulating Innovation in Products and Services: With Function Analysis and Mapping,
by J. Jerry Kaufman and Roy Woodhead
Copyright © 2006 John Wiley & Sons, Inc.

constrained within traditional five-day episodes but rather saw the approach as a key tool in an innovation toolkit: Berawi and Woodhead (2004), Bytheway (2004), Sato and Kaufman (2004), Seni (2004), and Woodhead and Berawi (2004).

7. Here are a few references to show a diversity of focus and underlying theoretical frameworks upon whose various approaches to particular value methodologies are based: Fallon (1980), Kaufman (1990, 1998), Kelly, et al. (2004), Miles (1961), Thiry (1997), Woodhead and McCuish (2002), and Zimmerman and Hart (1982). A bibliography can be found on the Veamac Web site (www.brookes.ac.uk/other/veamac), and SAVE International has a bookshop (http://www.value-eng.org/catalog_library.php) that contains many more titles than we can show.

8. The decision sciences are founded within a rationalist, and quite often in a hard positivist framework. As they deal with phenomena in an empiricist paradigm, they often ignore or fail to deal adequately with purpose, function, and value. This is not to suggest that their methods have no use, for they do. It is simply to state our view that they are methods that emerge from a narrow philosophical view of knowledge and human consciousness and are often caricatured as the logic of "bean counters" with narrow concepts of financial value confused with a more complete view of a valuable outcome. Because we have understood weaknesses in their underlying theories, we can now see the utility of their body of knowledge as contextual and so can treat them as tools selected for particular tasks. For us, their task is to seek out probabilistic evaluations of the way that things will turn out under various assumptions of uncertainty and then select what is believed to be the best choices available. For those wanting to know more about decision science we recommend Clemen (1992) as a good starting point.

9. The philosopher of science Karl Popper argued that no matter how many times a confirmatory observation had been achieved, there was always the possibility that a nonconfirmatory observation was available, if as yet undiscovered. He used the case "All swans are white," which until the discovery of black swans in Australia was thought to be an irrefutable truth. Popper wanted scientists to reach beyond their grasp and then set out to prove their own hypotheses wrong and in so doing reform them. For example, I might start with a hypothesis that "All water freezes at zero degrees Celsius." I would then search for evidence that would prove my hypothesis wrong. So when I discover that the salty seawater freezes at a temperature lower than zero degrees Celsius, I must revisit my original hypothesis and modify it so that it is again perceived as truthful; "All fresh water freezes at zero degrees Celsius" would be the next hypotheses I then try to disprove.

10. This relationship between object models, FAST models, and real-world artifacts is explained in Seni's (2004) paper.

11. Relativists believe that everything is relative to the observer. Some argue that if a tree falls in a forest where there are no people, it will not make a noise because the concept of noise requires that a human is involved. An extreme form of relativism is *solipsism*, where a person doubts the existence of everything, believing that his or her experience is simply a kind of dream. A more popular form of relativism is *skepticism*, in which objective truths are doubted, usually because they can be seen from different perspectives: One person might argue that a cup is half full whereas another can argue that it is half empty. Skeptics sometimes argue the case for multiple realities, as they see reality as existing in human minds. An example might be to consider the statement "Beauty is in the eye of the beholder," which implies that it is possible for everyone

to hold different notions as to what counts as beauty. What is ugly to one person could be beauty to another. We argue against these philosophical frameworks on the ground that they end up requiring power to seize the decision-making opportunity, whereas most people are paralyzed within clever semantics and sophism. Our position is drawn from the belief that if beauty is in the eye of the beholder, no single view of beauty is possible; we all hold different views. If that is the case, there can be no concept of beauty that we believe to be incorrect. For us, contradictory problems arise out of the relativist framework.

Postmodernism and *continental philosophy* build on such relativistic foundations. Again a contradiction exists that should cause us to question their founding premises. If we argue a case, relativists will counter that such contradictions are simply our opinions. Therefore, within the relativist framework, just as there can be no single concept of beauty, the concept of objective truth is also undermined. Our principal objection flows from a view of the contradictions. It seems that they put forward a singular view of reality that argues that there is no singular view. To them we exist in a state of multiple realities, so no singular reality can exist. To argue that there can be no single view (i.e., only multiple views) is a single view. Therefore, their entire agenda seems to be founded on a contradiction, so their argument seems to flow from an invalid premise.

In the fields of VE, VA, and VM such relativist views mean that advocates of postmodernist perspectives cannot see the relationship between a FAST model and reality. This is typically called *soft VM*. In their practice, FAST has little use other than as a means to stimulate wild ideas, and they invariably try to reduce the time invested in the construction of models, for they cannot grasp what Miles meant when he asked "What does it do?", for that is a question that looks beyond the physical part to an objective relationship in which a cog functions within a larger engine. Woodhead heard a VM practitioner in the UK claim that he could put a FAST model together in 15 minutes, as if the goal were to perform a task as quickly as possible. The underlying beliefs that make such actions seem like a good thing to do are based on an undervaluation of the worth of FAST modeling; the conversations are what enable innovation, and to cut them short misses the entire point.

12. The theme of nature being the outcome of underlying functionality was argued in a special edition of the *European Journal of Engineering Education* dedicated to value engineering. Woodhead and academic colleagues from the mobile phone industry argued that engineers mimic a deeper technology than the human-made version. Building on Ferré's (1995) philosophical examination of technology, we saw practical intelligence as technology and not the device itself, which performs mindlessly, as the designer intended. This led to a paper (Woodhead et al., 2004) that argued a dual technological ontology which allows us to see FAST as a tool to help us think beyond that which is observable and also to manage such conceptual insights and convert them into engineering technology.

13. Rationalists use logically coherent arguments in consistent ways founded on Descartes's "I think, therefore I am," which posits human consciousness as central to all other modes of questioning. (Other philosophers have argued against this, such as the continental philosopher Martin Heidegger.) That is, rationalism works from first principles to logical conclusions. If an initial theory proves to be wrong, the rationalist is forced to revisit the theory to once again achieve consistency between that which is predicted and that which is the result (a priori knowledge matching a

posteriori knowledge). They believe that everything is available to be known through methodologically consistent inquiry. A branch of rationalism that became known as *positivism* began with Comte, who argued for evidence to be used in all the sciences, including social physics, which he later named sociology. The term *positivism* was adapted by the Vienna Circle in the 1920s to go a step further by saying that only that which can be perceived either directly through our senses or indirectly via instruments counts as that which can be known. In this narrow sense, phenomena such as culture did not count as existing in a knowable way and therefore did not really exist. The physicists' views of what did or did not have merit dominated. If something could not be modeled mathematically, there was a good chance that they would not have allowed it to have any merit in their discourse. As with many schools, the rules pass into other contexts, such as social housing and traffic management, in which irrational humans become participating variables. In recent years the narrow view of positivism has been challenged and confined to laboratories, as its assumptions were seen as naive (e.g., the unsinkable HMS *Titanic*). However, the broader view of positivism, which contends that we need evidence to substantiate theories, is still needed even if seen as unfashionable. The key question is: How can you know when to accept or reject a piece of information as being significant?

14. In *social constructivism* a collection of people negotiate what they believe and then form theories that steer actions based on those beliefs. If we all agreed that the world was flat and behaved as if it were flat, we would be operating within a social construction. If we all believed that everyone at one time in the past believed that the world was flat, and because of that ignored the ancient Greeks, who worked out the circumference of the Earth, again we would be operating within a social construction. Physicians in the Middle Ages agreed among themselves that the "letting of blood" helped sick people to recover, so leeches were used extensively until another group of theories caused doctors to give less priority to their previous beliefs. Most markets are socially constructed: If I have a barrel of oil in my garage and last week the market offered $20 and this week they offer $40, this will influence my asking price should I want to sell the oil. No single person makes the market, however. It exists through the collective actions of buyers and sellers, as it is also socially constructed. To explore how we build socially constructed facts, Searle (1996) asks: What is the function of a coin when it is used as a screwdriver?

15. We are realists. *Realism* is a combination of rationalism and social constructivism. It accepts that some things exist that humans can never truly know or understand because they exist in ways beyond sensory or instrumental detection. This is the physics problem that string theory has, for strings are so small that they cannot be detected although we believe their existence to be possible. Therefore, there are two ontologies, one within the other. (An *ontology* is some class of things that exist independent of us.) There is the ontology of universal existence and the ontology of our awareness within it, which means that realism cannot be the same thing as the Vienna circle's narrow view of positivism (but could be similar to Comte's view of positivism). Our existence is constrained by the universal ontology, which functions in ways that we do not yet understand completely. For example, we know of gravity but don't really understand how it functions. String theorists believe that it is a force emanating from very weak particles called gravitons, but such is not a completely agreed social construction. Indeed, whether there is a single or multiple universes is another example of how

limited our knowledge is and how competing theories play a useful role in the trail-and-error building of knowledge through socially constructive systems of research and peer review processes. Realists see theoretical frameworks as instruments to help us manage our reasoning and believe that such frameworks are socially constructed. Realists must accept that assumptions play a role in directing which things are acknowledged and which ignored. Like blind men feeling our way through the dark, "truth" is thus a probabilistic measure of the accuracy with which a theory allows us to make reliable predictions of how reality will turn out to be in a future state. A FAST model is thus a realist construction that is built with people in the act of social construction but is referenced externally to allow a Comte-like approach to positivism to check our model before we analyze it to reason what needs to be done. Our human condition means that it can never be a true representation, so we must adopt a humble view of our knowledge.

16. Variance analysis considers the ramifications of planned outcomes versus actual outcomes. During the design stage the actual outcomes can be simulated using Monte Carlo.

CHAPTER 2

1. System dynamics models the observable outcomes of systems that interinteract. This is a tool that sits within a body of knowledge which helps those responsible for key decisions to simulate various changes and anticipate probable outcomes. For example, as customer complaints rise there will be a delay and then we would expect to see sales revenue fall. The weakness of system dynamics is that it does not delve into the functioning that leads to the outcomes it models. For example, we don't understand why customers are complaining, nor do we understand the functional relationships between the volume of complaints and the impact on sales volume; what has to be done to turn things around? Having said that, this discipline offers a useful way to start the strategic-problem-framing workshop and points us to areas and issues we need to think through in terms of functioning. Two important books will open the door to systemic thinking along the lines of system dynamics: Senge (1990) and a follow-on book that outlines various tools and techniques, Senge et al. (1994).

2. Chares Bytheway is a good friend who freely shares insights and wrote the Foreword to this book.

CHAPTER 3

1. Details of the value methodology standard can be found at http://www.value-eng.org/about_vmstandard.php.

2. Miles (1961, pp. 20–35).

3. We must remember that the way people slice things up so as to consider them is artificial. Reality is not sliced up, and the essentiality of functions is connected to a dynamically unfolding reality in which parts wear out, people get old, and the weather changes.

4. Those wishing to learn more about customer FAST are referred to Snodgrass and Kasi (1986).

5. The basic function of an oil filter relates to its work of capturing particles. We establish *Trap Particles* as the function. It's what it does that contributes to the system. If we see this as an intrinsic function, the outcome, purpose, or extrinsic function could be *Clean Oil*.

The basic function of a screwdriver is about the usefulness of such a device. A screwdriver achieves the driving of screws by doing what? It achieves this by enabling the turning motion of the hand to cause the screw to twist, and the screw's threads then cause the screw to enter the wood. So the function of a screw driver is *Transmit Torque*.

The basic function of a wall thermostat is what it does that enables room temperature to be controlled. If we said that the thermostat achieved that outcome directly, why would we need central heating systems? All that would be required would be thermostats. A thermostat does work within a system. It makes a valued contribution to that system. So our enquiry must look at what it actually does in such a way that we could think of alternative solutions. The fact is that a thermostat is really a kind of switch that turns the heating system on or off automatically depending on temperature. When cold, it forms a connection that allows an electrical current to flow that causes . . . . Its function is thus *Control Current*.

6. This is because reality has no scope lines and functions as a single entity. We exist within the existence of everything. As the Earth circumvents the Sun, a plane flies over the Atlantic, two lovers quarrel in Paris, cars pollute the environment, the sea erodes the shoreline, and so on. We cannot think in the same way that reality functions, so we slice it into chunks that our human capabilities are comfortable with, but we must see that just because that helps our ability to think, it does not mean that reality itself unfolds in this way.

When Descartes said "I think, therefore I am," he brought the illusion that reality follows from our consciousness. We argue the opposite and in this regard agree with Heidegger, who posited "I am, therefore I think," which positions existence before our consciousness is possible. However, we move beyond Heidegger's phenomenological inquiry into a realist one.

7. It was tempting to write "reduce the cost of the product here" but that might have misled. Value is increased through the optimization of solutions with respect to the outcome achieved by their combined permutations. The goal is to improve and make better. As efficiency is improved, cost invariably falls over the entire system but not necessarily for every function. We may reduce total cost by selecting an expensive solution for one function that enables almost zero-cost solutions for all the other functions. This is why customers control revenue in efficient markets; they evaluate the total offer placed in the marketplace. Myopic management often sees cost reduction as the goal and often degrades the offer placed into the marketplace. Once customers associate a company with poor products, that company will have little choice but to develop revenue-earning products in line with their poor brand image. We often seek elegant solutions that increase profitability rather than cheap solutions whose proponents fail to understand that customers control what succeeds or fails.

CHAPTER 4

1. The fundamental belief that reality is singular as opposed to plural is the key to understanding how FAST models attempt to link value and potential solutions. That

we humans may see reality differently (i.e., different perceptions and thus different interpretations) does not mean that reality is multiple; the issue is not with reality but with our interpretation of reality. The reason that we break things down into discrete parts is to help us think; analytical inquiry is in large part a product of our cognitive processes and capabilities. By considering parts of reality and how those parts connect locally to the larger whole aids our facilities to glean insights. These analytical insights are then mapped back into a connected and dynamic single reality within which our physical and conceptual being exists. This starting with the whole, breaking it down into parts for analysis, and then reconnecting to the whole in a synthesis is a principal theme in the highly recommended book by Carlos Fallon (1980). The following extract gives a flavor of the interconnectedness of value methodology built out of the realization of a synthesized reality. "The four most significant aspects of product value, from the standpoint of its analysis, are use value, esteem value, market value, [and] exchange value. We should not assume that by thus classifying them we are separating them from one another in real life. Analysis takes things apart to understand them. To function they have to go back together" Fallon (1980). (Details regarding this book are available from SAVE International's bookshop at www.value-eng.org.)

2. Charles Bytheway explained at a SAVE conference how he used a FAST model and arrived at a 50% value improvement. He did not accept this as the final outcome and repeated his use of the FAST model on this initial solution to find a further percentage improvement. Hungry for still more value, he repeated the study a few more times to get closer and closer to an elegant solution. It was only when subsequent attempts failed to enable value improvements that he realized that he must have found the best solution available. This aggressive search for innovation is an important aspect for successful FAST modeling, for although it is hard work, it leads to improvements in the concept stage that leverage annualized profitability. We recommend that everyone seeking to become better at FAST modeling search out papers and books by Bytheway.

3. In a blunt way, an organization faces one of three long-term strategic outcomes: (1) it will grow, (2) it will stabilize, or (3) it will shrink. The value realized through innovation will affect how an organization moves toward one of the three strategic outcomes. The FAST model must be linked conceptually to a consideration of these strategic outcomes so that the fruit of their production plays a beneficial role in shaping the destiny of an organization that takes investment risks in order to achieve rewards.

We are aware that in some industries, such as construction, the link between a project such as building an office block and the long-term strategic outcomes is disconnected. In such episodes, practitioners map the functions provided by the physical building by asking such questions as: What is the function of an external wall? Given that the function of one external wall will be similar to the function of another, it may be possible to develop some generic FAST models that are tweaked to fit a particular situation. If such is done, the practitioner could develop a FAST model that links the act of design and construction to the strategic outcomes that the client seeks. It may be possible for the two to be linked so that questions such as "How do we attract new customers?" could be held against "What type of external walls would influence new customers to select us as their providers?"

In the value methodology the pre-event stage is where the strategic outcomes are anticipated. The strategic problem and opportunity framing conversations link product improvement to organizational results precede the FAST modeling process. Kaufman (1998) wrote a paper that deals with this issue.

4. Best value is determined at the strategic or total system level and not at the individual solution level. The choice of a particular solution is made in terms of how it contributes to an elegant investment of ethics, time, energy, and money.

5. The expert FAST modeler always remains sensitive to influences that sometimes creep into the logic but not necessarily in a noticeable way. That is why the clear objectives and strategic context must precede the act of FAST modeling so that they reduce the opportunity for inadvertent drift. It is also why reflective practice and double-loop learning are so fundamentally important to the FAST modeler's own personal development. Three books that will help those meeting these concepts for the first time are Argyris (1977), Argyris and Schön (1974), and Schön (1983).

6. This was the subject of two papers that bring the nuances of the underlying theories of FAST modeling to the surface: Berawi and Woodhead (2004) and Woodhead and Berawi (2004).

7. Dan Seni of Quebec University at Montreal articulated the journey from the world of things into the world of ideas to produce a FAST model which was then innovated before producing a simulation to test theories and then on into prototyping (Seni, 2004). The paper provides well-received insights into the underlying theories that link FAST models to a wider innovation process. This paper was named "Paper of the Year" by SAVE International in 2005.

8. A couple of texts that explain customer FAST are Fowler (1990), Snodgrass and Kasi (1986) (contact SAVE International's bookshop at www.value-eng.org), and Thiry (1997).

9. However, the *if–then* and backward test "caused by" checks force a logical dependency in the vertical direction's imposition of causality on the horizontal intentionality. This crossing of causality and intentionality is at the root of practical ingenuity. The intentionality of *how–why* is what managers and engineers want. The vertical represents understanding drawn from science that itself includes no intentionality. The managers and engineers have a need and they search for things that will do that which is required.

10. The basic function we have selected may change as our model brings more informed insights.

11. That is, the internal perceptions of the team's thinking have been satisfied. This does not guarantee that they are complete or even correct. For greater reliability and confidence, the FAST model should be validated with external evidence.

12. The major logic path often links strategic and operational intentions. The vertical *when* or *if–then* brings the cold realities of causal factors that limit or condition the ease with which we seek to realize our intentions. Together they allow us to combine intentionality with causality.

13. Customer value is the key to an organization's medium- and long-term strategic outcomes. This is because customers control revenue in free and democratic markets.

14. The attempt at integrated thinking that unifies product and process is something FAST models enable as an innovation in that one nearly always ripples into the other, so to try and consider just one of them is to miss systemic opportunities.

15. This problem should always be traceable to the consequences that proposed solutions will have on long-term strategic outcomes of growth, stability, or decline.

This seeks to avoid inappropriate quick fixes that end up causing more problems. A clear line of sight should look to long-term survival.

16. An excellent introduction to systemic thinking and system dynamics can be found in a highly recommended book, *The Fifth Discipline*, by Peter Senge (1990).

17. By stepping back and looking at what we are doing here it can be seen that we are trying to glimpse a means–ends and cause–effect interpretation of the strategic context within which the study operates. Implicit in this realization is that this context, too, could be represented by a FAST model. Given that we are claiming the strategic context could be represented by both a FAST model and a system dynamics model, what binds the two approaches together?

System dynamics considers the relationships between empirical data. It maps the outputs of processes and how such processes cause or influence subsequent processes. The story "The Beer Game," which appears in Senge's (1990) book *The Fifth Disciple*, relates how changes in a system of connected processes go awry when each process is considered in isolation from their collective existence. At no time does it talk about the functionality of the beer and the role that can play other than indirectly through some pop group drinking the beer and enhancing brand value. So it's like putting a stethoscope to someone's heart and judging the functional health as a consequence of a process interpretation. For busy senior managers such tools allow a glimpse of the systemic context that is judged by them as sufficient to diagnose what strategic attributes need to be improved or degraded and which are more important than others.

The FAST model represents what is actually going on. So a FAST model of a human heart will explain how it works but lacks dimensions, such as the number of beats per minute, until we dimension it. There is a symbiotic relationship between FAST models and system dynamic models that has been combined in the value methodology to benefit the way they help us to understand what needs to be achieved and what needs to happen. Both are drawn out of a belief that reality does not stop and that snapshots of reality reduce the rich interpretations necessary for good management. For example, a CEO needs far more information to run a business successfully than appear in the balance sheet and profit and loss account; these simply tell him or her how things were on a single day.

18. Cost cutting is but one strategy; there are others, such as quality leadership. The one that is perceived as leading to a desired strategic outcome is the one that we should select. To decide on a strategy automatically without understanding the systemic forces at play in the strategic context is comparable to buying a lottery ticket. For some organizations, the decision to be the lowest-cost provider is based on an analysis of local competition which completely ignores the fact that some nations have very low labor costs that can never be beaten. Although their strategy might gain them a short-term cost advantage over local rivals, customers learn that they are low-cost providers, which limits their ability to reposition themselves. This "classical strategy" thinking is dealt with very well by Michael Porter (1980, 1985). Other approaches to strategy can be found in Sheldrake (2003) and Whittington (2001).

A supplementary bibliography to help those starting to read in this area of strategy and the organization follows: Andrews (1980), Ansoff (1965), Chandler (1962),etc. Cyert and March (1956), Drucker (1968), Fayol (1949), Follett (1924), Granovetter (1985), Hammer and Champy (1993), Herzberg (1966), James (1985), Janis (1982), Jarrillo (1993), Johnson and Scholes (1988), Lawrence and Lorsch (1967), Maslow (1943), Mayo (1933), McGregor (1960), Miller(1992), Mintzberg (1973), Pedler et al.

(1991), Peters (1988), Peters and Waterman (1982), Pfeffer and Salancik (1978), Quinn (1988), Schumacher (1993), Simon (1997), Slatter (1984), Taylor (1947), von Neumann and Morgernstern (1944), Williamson (1991), and Womack and Jones (1996).

19. Value management is an application of value methodology in soft or intangible processes such as management systems and decision making.

20. The pipe string is the metal rod that connects the drilling rig to the drill bit.

21. It is possible to use expandable pipe that is placed in situ, whose diameter is expanded by dragging a dilating plug through it. Subsequent pipe lengths are then fed through this enlarged diameter and the process is repeated. However, at the time of writing this book, this method was more expensive than the traditional methods and so was not considered the best choice.

CHAPTER 5

1. For further information about value methodology, read Fallon (1980), Kaufman (1990, 1998), and Woodhead and McCuish (2002).

2. In many construction projects, FAST models lose sight of this linkage. When Louis Sullivan said that "form follows function," he did not mean that we should focus our inquiry into the function of form in a way that detaches the strategic need for a building from its physical properties and their cost. The form is that which enables the function that yields value and gives purpose to the existence of the building. Together, the form that encloses functions and activities and the form that operates through its inherent functionality synthesize to achieve a value recognized after the act of building has been long forgotten.

3. This realization that a FAST model can help us to see what has to happen between key functions, combined with the view that a FAST model enables us to get closer to elegant solutions, shows a role for FAST prior to scheduling techniques such as PERT. To begin scheduling with techniques such as PERT requires us to have established what work needs to be done and in which order the work must be undertaken. PERT does not allow us to question whether activity x is vital and more often than not flows from a logic drawn from past experience. This observation is important, for the same FAST model can be used as the starting point for the coordination of management systems, such as organizational design, schedules, technology selection, sustainability, lean tolerances, six sigma, and so on. The reason why this is possible is because FAST models are about making explicit how we as a team think in practical ways to achieve desirable ends. Once our thinking is clear, we deploy resources in the real world to make those ideas into irrefutable phenomena.

4. The value methodology used here was value management.

5. Remember that the purpose of modeling functions is to bring clarity, so unnecessary complexity should be avoided.

6. A building is an unusual product, as people exist within it rather than having an external relationship to it, such as when someone holds a phone or a new hair dryer. The strategic context is thus about the ambitions of the people and organizations that will inhabit the building. The *how–why* logic path for a building should be about the strategic ambitions of the inhabiting organization and "form follows function." The function of elements such as the function of a staircase should be on minor logic

paths that show how they enable the strategic value to be achieved. This distinction between "form follows function" and the "function of form" must not be confused or overlooked. The former is what makes a building useful and valuable to users. The latter is about how the construction industry delivers its products efficiently.

7. In scenarios where high levels of subjectivity may exist, the role of "as is" and "should be" object models become more important. As Seni (2004) showed, the FAST model is drawn out of the critique of the object model. Object models can be system dynamic models or other means in which simulations can be run. We start with a situation in the real world and create an object model of it that can be tested. By comparing its predictions against reality, it becomes the focus of the FAST modeling exercise and is used later to test whether innovation has really been achieved.

8. The entire basis of costing is fraught with the same cost allocation problem. This is because cost and price are market-based data that are moderated by the value judgments being made within society. The way that a company is valued is also a judgment call, as is proven by the fact that profit and loss accounts and balance sheets simply take a snapshot based on the market worth of assets on a particular day which may bear little resemblance to the true cost (as the collapse of Enron bears out). When allocating cost, no matter in what context, there are judgment calls that lean closer to the subjective than the objective. Some people meeting the sensitivity matrix for the first time can be uncomfortable with the notion that a $100 cost is shared over, say, five functions in ways that are intuitive. Accountants are usually far more comfortable with such situations and regularly seek to establish corroborating evidence that will justify the assertions being made.

9. In some instances the capital cost (capex) and operating cost (opex) are brought to a net present value to reflect whole-life costs.

10. As customers control revenue, our task as purveyors of value is to offer products they prefer to their alternative purchase options. Managers may control costs as they make resource allocation decisions, but unless customers feed their revenue streams so that cash inflows are larger than cost outflows, the firm will not achieve profitability.

11. The relationship between value engineering and a value creation template has been explored by Woodhead and McCuish (2002).

12. Using *if–then* leads to the causal consequences and realization that if I do x, I also have to do y and z, which cost money to deliver, so the cost of x depends on the costs of y and z.

13. Given that customers control revenue, the market price is used as the starting point for new product development. A target cost is established that the design team has to achieve. The opposite approach is *cost to design*, in which a design is produced and then its manufacturing cost is determined. The danger with this approach is that the design may be far more expensive than the market is willing to pay, and for complex products such as new aircraft, the necessary cost reduction may not be easily accessible.

14. When a customer purchases a whole thing, the value is determined with respect to the usefulness of the whole thing. That parts and subsystems exist within the "thing" is irrelevant to the customer's sense of value. For example, if you were to consider purchasing a car, would you obsess over whether it had circuit boards made in the United States, Europe, the Far East, or Latin America? It is the synergy of the parts coming together to form something that is affordable and useful that creates the value. The company that controls the means to form that synthesis at the point of sale becomes

a very powerful node in a supply chain, especially if they understand and are capable of brand management. If a company knows what customers want and controls distribution channels, they can turn around to the supply network and treat the manufacturing outputs as commodities and thus drive down the prices and profits of the manufacturing industry that actually makes the products they assemble and sell. With the advent of the global economy, this scouring of the potential supply base gets easier and easier. To avoid being seen as producing low-value commodities, manufacturers such as Intel have set about their own brand promotion strategies so that customers ask if Intel components have been used.

If manufacturers do not understand the strategic worth of owning the point-of-value realization and do little to try to manage its power to drive down their prices, they may find that profits erode or even disappear overseas to low-cost labor economies where short-term relationships and "lowest bid wins" dominate management thinking.

15. Value is determined in the marketplace, where all the systems and solutions within the total product perform useful functions to form a complete offer to customers.

CHAPTER 6

1. For additional information on selecting and analyzing attributes, see Kaufman (1998, pp. 70–90).

2. Senge et al. (1994).

CHAPTER 7

1. The best introductory text we have seen for decision analysis is Clemen (1996).

2. Note that *Produce Heat* is something that follows from nature and is a phenomenon we simply have to deal with, as it is a consequence of a core technology, in this case a filament light bulb.

3. Ideation International is a company specializing in the application and practice of the TRIZ methodology. Their URL is http://www.ideationtriz.com/home.asp.

CHAPTER 8

1. Williamson (2000).

2. This logic plays a central role in Goldratt's (1990) theories.

3. Popper (1979).

4. The key concepts in the knowledge management community seem to consider knowledge (here we are talking of knowledge among humans that can be shared socially) as if it exists as a phenomenon divorced of action. (Check out various Web sites in your browser to see examples of knowledge that is really only information.) We hear of knowledge portals and knowledge repositories, which give the impression that knowledge is somehow disconnected from mental/thinking processes. It is as if knowledge can be stored in the same way that water can be stored. The term *tacit knowledge* is another example of words having their meaning modified by the knowledge management community. Polanyi (1958), a philosopher of science, coined

the term to describe a kind of knowledge that is elusive, hidden to us, inaccessible, and inexpressible—that which cannot be spoken of. Yet the knowledge management community holds the view that it is simply material waiting for articulation and give the impression that they mean embodied knowledge, such as the skills needed to catch a ball.

We see the term *tacit knowledge* used by knowledge management consultants as ill-defined or poorly structured know-how in a person's head. This class of knowledge needs to be articulated better and made clearer so that it can be communicated from one person to another as *explicit information*. We need to distinguish between types of knowledge so as to be able to manage what those differences can lead to. Nonaka (1991) gives an account of different types of knowledge and how they can be used to meld and transform. We interpret his types of knowledge exchange as (1) tacit knowledge to tacit knowledge through <u>socialization</u>, (2) explicit knowledge to explicit knowledge through the <u>integration</u> of information, (3) tacit knowledge to explicit knowledge through <u>codification</u> of informal systems into formal systems, and (4) explicit knowledge to tacit knowledge as formal systems become the new norms of practice through <u>internalization</u> as embodied knowledge.

5. Drucker (1969).

6. Ikujiro Nonaka coauthored with Hirotaka Takeuchi a valuable text in which they consider knowledge management and innovation (Nonaka and Takeuchi, 1995); see also Nonaka (1991).

7. Leonard (1995).

8. In an academic meeting, Woodhead heard the term *generalist* used as a put-down. When position power and promotional pecking orders are located around expertise, the ability to function as a multidisciplinary team becomes more difficult. In Plato's Academy and Aristotle's Forum no such artificial boundaries to knowledge seem to have distracted from a combined approach to education and advancement. Both generalist and specialist knowledge are symbiotic.

9. Allison (1971) showed that the Cuban missile crisis could be explained through a number of themes or frames of reference. Each explanation could be presented in isolation, but it was their combination that enabled a richer understanding of what went on. From this we see multiple perspectives offering a means to mediate a richer view of complex situations.

10. The laws of scarcity and demand are assumed to apply to knowledge. The logic seems to flow like this:

• If consultant X possesses vital knowledge that manager Y needs, the manager will be forced to pay the price that consultant X asks. This is true only if the cost of hiring consultant X is less than the reward that manager Y seeks and if no other consultant, at less cost, is available.

• Student Z observes consultant X and infers that success goes to consultant X because he has focused on developing specialist knowledge. Therefore, student Z seeks to adopt a similar career strategy, and universities trying to win students start designing courses that lead to a more rapid and systematic fragmentation of knowledge.

What is often not fully thought through is the knowledge role of manager Y. He has generalist knowledge that allows him to understand how to combine specialist knowledge and make use of it to create value. This involves as much reading and learning as are undertaken within a specialist deep but narrow education. Let us not confuse the generalist with a lazy intellectual; generalists are people who have a systemic view of education and so inquire into many subjects. A problem that seems to be looming in the UK and the United States, especially among small and medium-sized enterprises (SMEs), is a shortage of capable generalists in influential roles. Industries such as manufacturing seem to be eroding under the low-cost advantages offered by countries such as China, because leaders in SMEs have been sheltered from strategic learning over the past 30 to 50 years. Protected from the need to develop strategic leadership skills through their slavelike relationship with a mass producer, such companies are often managerially unprepared for the need to be far more creative than they have ever been. The emphasis on narrow "expertise" has led many people to avoid developing generalist knowledge. It is now, in a period of industrial revolution that we call the *New Economy*, that many former specialists, scientists, and engineers rise to the top of a particular SME only to discover that there are far more variables at play than they have been prepared to deal with. The way forward is often seen as bringing generalist knowledge in via management consultancy; however, these people are often specialists that have simply focused expertise around business management processes and interpersonal skills; they have a different kind of I knowledge, so the T is still fragmented (I and T knowledge refer to Dorothy Leonard's work). What is needed is a way to unify the resident knowledge in the organization. Next, it needs to be assessed in terms of competency with respect to the capability to deliver services and products that yield profit.

11. Lesser and Storck (2001).

12. One community of practice (CoP) seeing itself as more important and more relevant to another will differentiate, and it is that which leads to compartmentalization. As such a "department" may be transformed into new compartmentalized structures based around specialist knowledge and expertise, so the fragmentation will continue within a different structure—the same problem trapped inside a different structure.

13. Social capital is the value of social relationships. Woolcock (2001) provides a useful starting point for further reading.

14. Here we are talking about single- and double-loop learning, with particular reference to the work of Chris Argyris (1991). The distinction between single- and double-loop learning is also discussed in Peter Senge's (1990) book, *The Fifth Discipline*.

15. Leonard (1995).

16. We also make reference to Aristotle's five ways of grasping the truth:

- *Epistémé.* This is a scientific perception of the nature of the world and how it functions or operates. This leads to the articulation of scientific laws such as Ohm's law or Boyle's law.
- *Techné.* Here craft-based know-how is considered alongside epistémé. This is a practical knowledge and ability to make things or change things that develops out of a relationship with tools, techniques, and materials. It is through repetition and exposure to situations that test the know-how that leads to embodied knowledge.
- *Phronesis.* This is an intelligence that sees opportunity flowing from techné. It is phronesis that sees an alternative way of achieving some useful outcome that is a kind of practical ingenuity. As such it is different from both epistémé and techné and sometimes is a product of their combination which leads to invention. Phronesis is action oriented and seeks to change things in

order to create new things. An entrepreneur such as Henry Ford demonstrated phronesis when he established mass production techniques back in the 1910s.

- *Nous*. Those having nous understand first principles. For example, a scientist with nous extends his or her inquiry beyond Boyle's observable fact that "for a fixed mass of an ideal gas at constant temperature, the pressure is inversely proportional to the volume." The scientist with nous inquires more deeply why that is true. A key issue here, then, is how to relate laws of science to laws of nature and whether such concepts as "laws" are valid. For the person with craft-based know-how, those with nous will figure out why things have not worked out as planned by looking to *first principles* and why first principles have credence. When Henry Ford started buying all the companies in his supply chain so that he could more easily control costs and quality, he was demonstrating nous.

- *Sophia*. This word means "wisdom." It is wisdom that blends epistémé, techné, phronesis, and nous to realize a capability to acquire advantage. It is important to understand that Aristotle saw everything as being connected. As such, talk of "knowledge" as a quasi-physical asset that can be managed could be challenged on the grounds that it also yields the possibility that knowledge can be separated from wisdom. In modern views it is possible to have someone who is knowledgeable but unwise. In an Aristotelian perspective, wisdom and knowledge cannot be separated and each is dependent on the other; if you have not much wisdom, you can't be all that knowledgeable, and if you are not that knowledgeable, you can't be all that wise. Let us be clear that those who have risen to the top in history have not always been wise, so there is an element of luck at play. Take the case of Henry Ford. We have argued above that he showed phronesis and nous when he enabled mass production and bought all the suppliers, so that he controlled all the supply chain. However, he nearly bankrupted the Ford Motor Company in the 1930s because he would not delegate, and as the enterprise grew he became a major bottleneck to decision making. In fact, if it were not for his son, Henry Ford Jr., the great Ford Motor Company may have slipped into the annuls of history like so may other pioneers who failed to stay at the forefront of progress.

What we are doing is trying to articulate a richer view of management concepts so that our ability to master events is extended. We can see that there is an interplay of different types of knowledge that may not fit comfortably within an organizational chart with the CEO at the top and craftsmen replaced by outsourced subcontractors at the bottom. Such a view of an organizational pyramid is often a representation of power but not necessarily knowledge or the means to access insights already in the heads of employees. For us, innovation needs to dismantle the pyramid and have information shared so that innovation is possible. The reality is that in the day-to-day running of an organization, the pyramid structure is useful. We therefore need the ability to have more than one organizational design: one to be most effective in terms of day-to-day operation and the other to encourage innovation.

17. See note 14.

18. This was achieved in part on a project for BP where two rival bids to build a North Sea production platform both exceeded base-case economics. We got BP and the two contractors to collaborate in building a FAST model. In this act the contractors learned, with the client, more about what was actually needed. A few key specifications

were revealed as being too vague to enable better pricing, which forced BP to firm up some decisions they had been delaying in order to enjoy market benefits later; BP had not realized that this uncertainty caused contractors to overprice certain items. Once a common view of the functions to be performed were agreed to, the two contractors ceased collaborating and then went back to their individual design offices to reinvent a new solution that would be better than that of the other contractor. This case study reveals that it is possible to gain benefits from a FAST model without necessarily giving away any secrets related to specific solutions or particular ways of performing functions. It also shows how a procurement route can be designed to contain both collaboration and competition.

19. Hall et al. (1994). These authors point out some of the failings of business process reengineering and suggest that almost two-thirds of BPR interventions have failed.

20. Here we acknowledge the insights provided by ex-Shell mangers, in particular Andrew Garnett of PTC Consulting Limited in Sutton Benger, Chippenham, Wiltshire, England. Garnett and PTC Consulting helped us to bring the theories presented by Chris Agyris, Peter Senge, and Bill Torbert to life in our action research. It is this series of conversations that helped us reinvent value engineering as *rapid innovation*, in which the practical knowledge in the heads of employees is managed rather than single-loop learning logic, focused on people, schedules, budgets, and specifications.

21. Italic type signifies that these are verb–noun functions that one might see on the horizontal *how–why* logic path of a FAST model.

22. If we accept that the ultimate decision as to which data we should believe are real and which illusory as a foundation for one's personal belief system, we have to accept that as people acquire greater knowledge and awareness, they may modify their original beliefs. This is the basis of any education system. If we accept this premise, we have to accept that not all people will necessarily hold the same personal beliefs throughout their lifetimes; change is enabled by learning. Different people can hold different personal beliefs at the same moment in time, and in addition, at a future time each of these people may have adapted or completely changed his or her original personal beliefs. If we accept that our psychology is built on founding personal beliefs and that these can change, we have to accept personality is dynamic. This forces us to question the usefulness of such tools as psychometric tests, as they are grounded in a theory of static personality and thus are not good predictors of future ability. Therefore, the key to changing behaviors in people is accessed through the core beliefs that steer their inner values and hence their observable actions. Clare W. Graves gave us a taxonomy that explores inner coping systems and how such systems make their owner reasonably predictable and potentially able to be transformed through education. For more information about Clare Graves and spiral dynamics, visit http://www.clarewgraves.com/home.html.

23. Between these points of skepticism and gullibility are other philosophical theories, such as materialism, determinism, phenomenology, relativism, positivism, and realism. It would be nice if we could tell you which one is "correct," but the truth is that we cannot do so irrefutably, as we, too, have made a fundamental choice as to how we believe mind and world relate; we are therefore suspicious of any group which proclaims that they have "the" one way. If we had to nail our colors to the mast, we will choose to be realists, believing that there is a single reality too complex for us to comprehend through our dulled senses and so inhabit a socially constructed

view of reality that itself functions through socially constructed systems such as the economy. For us, reasoning reigns supreme. The key word here, though, is *choose*, as we are fully aware that our belief system is founded on a single choice which has then steered our values and how we view priorities. This worldview necessitates that we all empathize with each other as we try to share insights and determine ways of proving our theories as learners.

24. Philosophers of technology argue that we are entering an era where humans and machines will work together more closely than ever before. Don Ihde (2002) concludes one of his books: "We are our bodies—but in that very basic notion one also discovers that our bodies have an amazing plasticity and polymorphism that is often brought out precisely in our relations with technologies. We are bodies in technology."

25. The observable causal relationships can be modeled with system dynamics software so that simulations can be run to see how the model correlates to real-world data. This can be used in parallel to a FAST model so that the effects of swapping one mode (i.e., solution) of performing a function for another can be tested in a simulation model. Go to the following Web site for examples of the software that can be used: http://www.iseesystems.com/(phv2n245yrjote45tpas0ryc)/Community/STArticles/SystemsThinking.aspx.

REFERENCES

Akiyama, K. (1991), *Function Analysis: Systematic Improvement of Quality and Performance*, Productivity Press, Cambridge.

Allison, G. T. (1971), *Essence of Decision: Explaining the Cuban Missile Crisis*, Little, Brown, Boston.

Andrews, K. R. (1980), *The Concept of Corporate Strategy*, Richard D. Irwin, Homewood, IL.

Ansoff, H. I. (1965), *Corporate Strategy*, McGraw-Hill, New York.

Argyris, C. (1977), Double loop learning in organisations, *Harvard Business Review*, September–October, pp. 115–125.

Argyris, C. (1991), Teaching smart people how to learn, in *Harvard Business Review on Knowledge Management*, Harvard Business School Press, Boston.

Argyris, C., and Schön, D. A. (1974), *Theory in Practice: Increasing Professional Effectiveness*, Jossey-Bass, San Francisco, CA.

Berawi, M. A., and Woodhead, R. M. (2004), A teleological explanation of the major logic path in classic FAST, *Proceedings of the 44th SAVE International Annual Conference*, Montreal, Quebec, Canada.

Bytheway, C. W. (1968), Simplifying complex mechanisms during research and development, *Proceedings of the Annual SAVE International Conference*, Vol. III, Washington, DC.

Bytheway, C. W. (2004), The genesis of FAST, *Proceedings of the 44th SAVE International Conference*, Montreal, Quebec, Canada.

Chandler, A. D., Jr. (1962), *Strategy and Structure: Chapters in the History of the American Industrial Enterprise*, MIT Press, Cambridge, MA.

Clemen, R. T. (1996), *Making Hard Decisions*, 2nd ed., Duxbury Press, Belmont, CA.

Cummins, R. (1975), Functional analysis, *Journal of Philosophy*, Vol. 72, No. 20, pp. 741–765.

Cyert, R. M., and March, J. G. (1956), Organisational factors in the theory of monopoly, *Quarterly Journal of Economics*, Vol. 70, No. 1, pp. 44–64.

Stimulating Innovation in Products and Services: With Function Analysis and Mapping,
by J. Jerry Kaufman and Roy Woodhead
Copyright © 2006 John Wiley & Sons, Inc.

Drucker, P. F. (1968), *The Practice of Management*, Pan Piper, London.

Drucker, P. F. (1969), *The Age of Discontinuity: Guidelines to Our Changing Society*, Harper & Row, New York.

Fallon, C. (1980), *Value Analysis, USA*, Value Foundation, Washington, DC.

Fayol, H. (1949), *General and Industrial Management*, Pitman, London.

Ferré, F. (1995), *Philosophy of Technology, USA*, University of Georgia Press, Athens, GA.

Follett, M. P. (1924), *Creative Experience*, Longmans, Green, London; republished in Sheldrake (2003).

Fowler, T. C. (1990), *Value Analysis in Design*, Van Nostrand Reinhold, New York.

Goldratt, E. M. (1990), *The Haystack Syndrome: Sifting Information Out of the Data Ocean*, North River Press, Great Barrington, MA.

Granovetter, M. (1985), Economic action and social structure: the problem of embeddedness, *American Journal of Sociology*, Vol. 91, No. 3, pp. 481–510.

Hall, E. A., Rosenthal, J., and Wade, J. (1994), How to make reengineering really work, *McKinsey Quarterly*, No. 2, pp. 107–128.

Hammer, M., and Champy, J. (1993), *Reengineering the Corporation: A Manifesto for Business Revolution*, HarperCollins, New York.

Herzberg, F. (1966), *Work and the Nature of Man*, Crosby Lockwood Staples, London.

Ihde, D. (2002), *Bodies in Technology*, University of Minnesota Press, Minneapolis, MN.

James, B. G. (1985), *Business Wargames*, Penguin Books, Harmondsworth, Middlesex, England.

Janis, I. L. (1982), *Groupthink*, 2nd ed., Houghton Mifflin, Boston.

Jarrillo, C. J. (1993), *Strategic Networks: Creating the Borderless Organisation*, Butterworth–Heinemann, Oxford.

Johnson, G., and Scholes, K. (1988), *Exploring Corporate Strategy*, 2nd ed., Prentice Hall, London.

Kaufman, J. J. (1990), *Value Engineering for the Practitioner*, North Carolina State University Press, Raleigh, NC.

Kaufman, J. J. (1998), *Value Management: Creating Competitive Advantage*, Crisp Management Library, Washington, DC.

Kelly, J., Male, S., and Drummond, G. (2004), *Value Management of Construction Projects*, Blackwell, Oxford.

Kroes, P. (1998), Technological explanations: the relation between structure and function of technological objects, *Techné: Journal of the Society for Philosophy and Technology*, Vol. 3, No. 3, Spring.

Lawrence, P. R., and Lorsch, J. W. (1967), *Organisations and Environment*, Harvard Graduate School of Business Administration, Cambridge, MA.

Leonard, D. (1995), *Wellsprings of Knowledge: Building and Sustaining the Sources of Innovation*, Harvard Business School Press, Boston.

Lesser, E. L., and Storck, J. (2001), Communities of practice and organizational performance: knowledge management, *IBM Systems Journal*, Vol. 40, No. 4.

Mahner, M., and Bunge, M. (2001), Function and functionalism: a synthetic perspective, *Philosophy of Science*, Vol. 68, pp. 75–94.

Maslow, A. (1943), A theory of human motivation, *Psychological Review*, Vol. 50, pp. 370–396.

Mayo, E. (1933), *The Human Problems of an Industrial Civilization*, Viking Press, New York.

McGregor, D. (1960), *The Human Side of Enterprise*, Penguin Books, Harmondsworth, Middlesex, England.

Miles, L. D. (1961), *Techniques of Value Engineering and Value Analysis*, McGraw-Hill, New York.

Miller, D. (1992), The Icarus paradox: how exceptional companies bring about their own downfall, *Business Horizons*, January–February, pp. 24–35.

Mintzberg, H. (1973), *The Nature of Managerial Work*, Harper & Row, New York.

Nonaka, I. (1991), The knowledge creating company, in *Harvard Business Review on Knowledge Management*, Harvard Business School Press, Boston.

Nonaka, I., and Takeuchi, H. (1995), *The Knowledge Creating Company*, Oxford University Press, Oxford.

O'Brien, J. J. (1987), *Lawrence D. Miles Recollections*, Miles Value Foundation, Washington, DC.

Pedler, M., Burgoyne, J., and Boydell, T. (1991), *The Learning Company: A Strategy for Sustainable Development*, McGraw-Hill, London.

Peters, T. J. (1988), *Thriving on Chaos*, Macmillan, New York.

Peters, T. J. and Waterman, R. H., Jr. (1982), In *Search of Excellence*, Harper & Row, New York.

Pfeffer, J., and Salancik, G. R. (1978), *The External Control of Organisations: A Resource Dependence Perspective*, Harper & Row, New York.

Polanyi, M. (1958), *Personal Knowledge: Towards a Post-critical Philosophy*, University of Chicago, Chicago.

Popper, K. R. (1979), *Objective Knowledge: An Evolutionary Approach*, Oxford University Press, Oxford.

Porter, M. E. (1985), *Competitive Strategy*, Free Press, New York.

Preston, B. (1998), Why is a wing like a spoon? A pluralist theory of function, *Journal of Philosophy*, Vol. 95, pp. 215–254.

Quinn, R. E. (1988), *Beyond Rational Management*, Jossey-Bass, San Francisco, CA.

Sato, Y., and Kaufman, J. J. (2004), VA tear down: a new value analysis process, *Proceedings of the 44th Annual SAVE International Conference*, Montreal, Quebec, Canada.

Schön, D. A. (1983), *The Reflective Practitioner*, Basic Books, New York.

Schumacher, E. F. (1993), *Small Is Beautiful: A Study of Economics as if People Mattered*, Vintage Books, London.

Searle, J. R. (1996), *The Construction of Social Reality*, Penguin Books, London.

Senge, P. M. (1990), *The Fifth Discipline: The Art and Practice of the Learning Organisation*, Century Business, London.

Senge, P. M., Kleiner, A., Roberts, C., Ross, R. B., and Smith, B. J. (1994), *The Fifth Discipline Fieldbook*, Nicholas Brealey Publishing, London.

Seni, D. A. (2004), Function analysis as a general design discipline, *Proceedings of the 44th SAVE International Annual Conference*, Montreal, Quebec, Canada.

Sheldrake, J. (2003), *Management Theory*, 2nd ed., Thomson Learning, London.

Simon, H. A. (1997), *Administrative Behaviour*, Free Press, New York.

Slatter, S. (1984), *Corporate Recovery*, Penguin Books, Harmondsworth, Middlesex, England.

Snodgrass, T., and Kasi, M. (1986), *Function Analysis: The Stepping Stones to Good Value*, University of Wisconsin, Madison, WI.

Taylor, F. W. (1947), *Scientific Management*, Harper & Row, New York.

Thiry, M. (1997), *Value Management Practice*, Project Management Institute Publications, Sylva, NC.

von Neumann, J., and Morgernstern, O. (1944), *The Theory of Games and Economic Behaviour*, Princeton University Press, Princeton, NJ.

Whittington, R. (2001), *What Is Strategy and Does It Matter?* Thomson Learning, London.

Williamson, T. (1991), Strategizing, economizing and economic organization, *Strategic Management Journal*, Vol. 20, No. 12, pp. 107–108.

Williamson, T. (2000), *Knowledge and Its Limits*, Oxford University Press, Oxford.

Womack, J., and Jones, D. (1996), *Lean Thinking: Banish Waste and Create Wealth in Your Corporations*, Simon & Schuster, London.

Woodhead, R. M., and Berawi, M. A. (2004), An etiological explanation of WHEN logic in classic FAST, *Proceedings of the 44th Annual SAVE International Conference*, Montreal, Quebec, Canada.

Woodhead, R. M., and McCuish, J. (2002), *Achieving Result: How to Create Value*, Thomas Telford, London.

Woodhead, R. M., Ball, F., and Li, X. (2004), Silk flowers are just as artificial as plastic ones: value engineering in the university context, *European Journal of Engineering Education*, Vol. 29, No. 3, September, pp. 333–341.

Woolcock, M. (2001), The place of social capital in understanding social and economic outcomes, *Canadian Journal of Policy Research*, Vol. 2, No. 1, Spring; http://www.isuma.net/v02n01/index_e.shtml.

Wright, L. (1973), Functions, *Philosophical Review*, Vol. 7, pp. 139–168; reprinted in Buller, D.J., ed. (1999), *Function, Selection and Design*, State University of New York Press, Albany, NY.

Zimmerman, L. W., and Hart, G. D. (1982), *Value Engineering: A Practical Approach for Owners, Designers and Contractors*, Van Nostrand Reinhold, New York.

APPENDIX: FREQUENTLY ASKED QUESTIONS

The following are answers to questions asked frequently following the completion of a learning course and workshop using function analysis system techniques (FAST) modeling. The way we have written the responses assumes that you have read the book. As such, we do not spend time defining terms such as *basic function*. If you have not read the book, our answers may be difficult to grasp on first reading. We advise you to read the book before reading these question and answers.

Q. Three questions are asked prior to starting *a* FAST model; why are they asked?

A. The questions to which you are referring are:

- What is the problem (or opportunity) that we are about to resolve?
- Why do you consider this a problem (or opportunity)?
- Why do you believe that a solution (resolution) is necessary?

The questions are asked at the opening of a VE assignment to ensure that all team members are addressing the same problem and to uncover any hidden agendas and systemic trends of which we should be aware. Essentially, the questions are meant to get us as a team to explore the context within which our interventions will be judged. It is from such strategic opportunity and problem framing that we gain a clearer view of both value and success and who will evaluate it. The questions are also asked when FAST is used to map a new product or process functionally. Contained within the answers to the first question are guidelines to the FAST model's primary theme and level of abstraction. The second and third questions, starting with *why*, test the first question to ensure that we have

Stimulating Innovation in Products and Services: With Function Analysis and Mapping,
by J. Jerry Kaufman and Roy Woodhead
Copyright © 2006 John Wiley & Sons, Inc.

separated the problem from its symptoms. The answer to the first *why* looks to a higher level of abstraction. The second *why* tests the two preceding answers and seeks to build a view of the benefits or lack of benefits that might make solving the problems worthwhile.

Q. Why create a FAST model, and where is it taking us?

A. Let's drop back a moment and agree that function analysis is the foundation of the value discipline. It is the one approach that separated us from a myriad of cost reduction initiatives. FAST is one of many methods to analyze function. Others include random function determination, the function tree, numerical function analysis, function hierarchical, and the function–cost matrix, to name a few; all are good and each is effective if used properly. The FAST model has the advantage of using logic to represent the necessary contributions in a dynamic system. The key point is that a team is forced to discuss why various things are needed in a way that obviates preconceived solutions. This builds a deep and rich understanding as a team rather than having just one person understanding and hence controlling. If a function does not fit in the logic of the model, it could be that the function was not described properly; here the team is again forced to talk across departmental and disciplinary boundaries. In some cases it is likely that the verb must be replaced because the *how* and *why* questions that the verb tops build a story of intentionality. That's also why the verb used to describe a function should be active rather than passive. A missing function or a function placed too high or too low on the level of abstraction can also cause the model to be read incorrectly. Our goal is to build a reliable representation of reality that helps us to innovate. We must also be aware of approaches that are under the influence of a dominant personality in the team who decides how to describe functions and whether the functions are basic or secondary. Not only will they run the risk of guiding the conversations but may degrade the link between the FAST model and the reality it purports to represent. FAST is a communication tool that heightens understanding and empathy in such a way that practical ideas flourish. As such, the model itself is not as important as the discussions it stimulates among the team members in attempting to comply with the rules governing the model-building process and how such verb–nouns logically linked relate to the working reality they know. You will find that the process forces an in-depth discussion of how things work, their functions, and their dependencies on other functions. Once the model is complete, it can be dimensioned in a variety of ways to establish goals and objectives, design-to-cost models, budget justifications, speculation topics, and so on. One project involved a hospital's operating room effectiveness. The six participating departments framed their department's FAST models and used them to determine which function was nonresponsive when a problem surfaced and which dependent (input) functions might be responsible. The team also used their models to explain to visitors how their department functioned.

So a FAST model operates on many levels and is useful because the act of building it forces teams to think clearly about what is necessary in a way that encourages many new ideas. Without it, a team may well have generated many new ideas, but how could we be sure that the ideas would affect the value? By virtue of the logic rules in FAST modeling the link between a particular solution and value is a product of clear lines of sight from the sensitivity matrix through FAST and the paired comparison to the range of goodness.

Q. How does function analysis relate to FAST?

A. Function analysis describes a family of techniques used to analyze functions, including FAST. However, FAST stands out because the modeling processes clearly illustrate function dependencies. The function of a component is only valuable as part of a working system. That system works because another component has at least one function and so is dependent on the selected part. Ideally, we could create a new component that performs both functions. FAST models make the relationships between one function and another explicit. In traditional function analysis such interdependencies are implied but not made explicit.

Q. How do you apply cost and worth to FAST models?

A. Once a FAST model is complete, the functions can be clustered or otherwise dimensioned to display the part costs to perform the functions. On a portable compressor, for example, the air end, compressor, running gear, and so on, were displayed with their actual and target costs and then broken down further to components within the subsystems. The same process applies to an organization or procedure, where actual cost against budget can be assigned to departments and the functions within departments. As to "worth," we believe that the market determines the value of the product or service offered but that the buyer or customer determines its worth. As an example, if a producer prices a product at $100 and you purchase the product, you did so because you thought the product was "worth it." In this regard, worth has nothing to do with cost. The fact that a product is expensive to produce does not influence the customer to buy the product. When we talk about customer perceived value, we are referring to the producer's analysis of what functions and attributes will attract and influence the customer to purchase the offering. If the customer turns away from the sale, it is because the customer's sense of "worth" did not equal the value (price) asked by the producer. If you are looking for a more specific definition of worth, I refer you to VE's definition: "Worth is the lowest cost to perform the function reliably." So cost relates to real-world solutions such as the amount of money spent to buy a pound of nails. Worth relates to the way that purchased solutions perform other functions, such as nailing down a roof tile that keeps you awake at night when the wind blows and the money spent buying nails so that you get a good night's sleep.

Q. Is it necessary to do an "as is" model prior to a "should be" model?

A. No. If you are doing a product improvement FAST model, an "as is" model is suggested. This allows you to analyze the way that the functions are currently being performed and (if appropriate) the cost to perform those functions. Once the project has been completed successfully, there would seen to be no reason to reconfigure the FAST model to reflect the changes unless it will help with your closing presentation. A "should be" model is used to describe a new concept functionally when there is no product template. It is easier to construct a "should be" model, because in configuring an "as is" model, all the components must be represented in the model functionally to arrive at a cost–function assessment. A "should be" model has no components; it is conceptual.

Q. What is the correct step-by-step procedure to follow in creating a FAST model?

A. A complete answer would require that you reread the book, but let us offer some hints. All FAST models, whether simple or complex, hardware or procedural, should start by listing those key random functions that best describe the theme or project's primary objective. After identifying from 10 to 15 random functions, expand them in the *how* and *why* directions. This will give you 30 to 45 related random functions use in beginning to build your FAST model. Other functions are added as you build the logic paths. When you believe you have configured the major logic path, identify the basic function(s) and input function(s), which will locate the scope lines. Then place the *when* functions, those functions that link to the major logic path functions. An important point to remember is that a FAST model is built by a team and must have team consensus to be valid. This validity is a product of the team's knowledge about how the model relates to the real world and the laws of both science and nature.

Q. What is the best way to teach FAST?

A. The best approach that we have found is to involve students in the FAST building process as soon as possible rather than confounding them with principles, rules, regulations, variations, and applications. After students have some understanding of how to use the verb–noun syntax and the difference between basic and secondary (supporting) functions, explain the *how–why* dependency concept, scope lines, and *when* and *if–then* operands. Then, using a simple mechanical example, lead the students as they identify the random function process and encourage them to begin building the model. Remind students that Post-It notes can be moved around easily to learn and test function dependency. This introduction takes about 30 minutes. During the event, students will question why one approach is better than another, and why you rejected or recommended a different way of configuring the model. This is the opportunity to relate the principles, methods, and techniques of FAST modeling to a receptive audience. The more the students work on case study examples, the greater will be their knowledge and appreciation of the FAST methodology.

There is a paradox in using hardware examples to teach FAST. Although students can relate their model to the hardware example easier than to soft or new product development activities, existing hardware represents an after-the-fact "as is" model. Therefore, when a student's model disagrees with the way a hardware example is configured, his or her first thought is that the FAST model must be wrong. The fact that the reverse may be true, with the FAST model pointing to a fault in the example product, is rarely considered. We have tried teaching the FAST discipline by first using concept examples where no hardware existed, but new students need something visual to which to relate their FAST model.

INDEX

Abstract function, 95
Abstraction level:
  basic function rules, 52
  determination of, 175
  hierarchical techniques, 39, 43
  high, 109, 177
  implications of, generally, 9–11, 40
  influential factors, 94
  low, 114, 157–158
  random functions, 54–55, 58
  sample model-building process, 85, 88
  selective lower level, 104
Abstract nouns, 67
Accountability, 27, 95, 140
Active verbs, 45–46, 48, 66–67, 79, 84, 89
Administrative process value studies, 96
Aesthetic functions, 50, 56
Airlifting objects, problem-solving case illustration, 39–41
Alternative solutions, 37, 46, 155, 162, 196
Ambiguity, 27–28, 32–33, 46, 50, 75, 79
Amortization process, 106, 109
And-or questions, 73–75
And relationships, 73–75
Aristotle, 213 n8, 214–215 ns16–17
Artifact analysis, 104, 106, 114
"As is" model, 62–63, 70, 94, 103, 106, 111, 211 n7, 224
Attributes:
  acceptance range, 118–120
  boundary conditions, 134–135
  construction management case study, 128–138
  identification of, 115–118
  incentives for FAST model, 138, 144–149
  incorporation into FAST model, 120–121
  influence on FAST model, 138
  linked issues to FAST model, 121–128
  performance curves, 144
  ranking system, 120, 145
  selection of, 166
  significance of, 20, 115
  software acquisition case study, 138–142
  validity of FAST model, 140, 143–149
  weighting, 136–137, 145–146
Automotive parts case study, 191–193

Basic function(s):
  characterized, 206 n5
  defined, 5–6, 38, 49
  hierarchical technique, 39–40
  identification, 223
  implications of, 13, 58, 60, 69–70
  influential factors, 94
  innovation points, 162
  random function determination, 55
  sample model-building process, 77–78, 85
Belief system, 216 n122
Bending the rules, 87–88
Beneficial, 15, 22

Best-in-class goals, 114, 116
Bias, 186
   implications of, 186, 196
   problem-solving techniques, 29, 39
   in problem statements, 160
Blue-sky research, 161
Body language, 178
Brainstorming, 12, 70, 169, 173–174
Breakthrough innovation, 28, 71
Budgeting, 95, 110–111
Business metrics, 183
Business process reengineering, 34
Bytheway, Charles, 13, 38, 205 n2, 207 n2

Capital equipment, 95, 104
Case-specific dimensioning, 93, 109
Causality, 66–67, 71, 179, 186
Causal models, 18
Causal relationships, 59, 61–62, 123–125, 131,
   152, 195
Cause-and-effect relationships, 71, 121
Chain of command, 184
Chains of functions, 61, 71
Change process, 16, 29, 155, 162
Charles's law, 180
Charter, 86, 111
Checklist cultures, 23
Chief operating officer (COO), 186
Cigarette lighter, case illustration:
   basic function rules, 51–52
   FAST modeling process, generally, 61
   hierarchical technique, 38
   innovation case illustration, 162
   sample model-building process, 76–84
   verb-noun function technique, 29–30
Class, problem-solving techniques, 30
"Clean sheet" approach, 103, 139
Clustering:
   attributes and, 121
   benefits of, 104, 196, 224
   functions, 111–114
Co-development, 169
Coercion, 22
Collaborative innovation, 60
Collaborative inquiry, 193
Collective intelligence, 62
Collective knowledge, 186
Colloquial expression, 6
Color coding, 104, 193
Commitment, significance of, 32, 186, 189
Communication(s), generally:
   applications, 9
   clarity of, 62, 189
   competent FAST modelers, 176
   importance of, 92
   interdisciplinary teams, 16–17
   jargon, 60

   two-word function descriptions, 47–48, 89
   within team, 15–16
Communities of practice (CoPs), 182, 214 n12
Comparative relationships, 106–108
Competent FAST modelers, 176–177
Competitive advantage, 187, 189–191
Competitive benchmarking, 193
Competitive position, 29, 86
Completion dates, 99
Computer projector FAST model, 156–157
Comte, 204 n15
Concept design, 19–20
Conceptualization, 23
Conceptual thinking, 45
Concurrent engineering (CE) projects, 111
Consensus, *see* Team consensus
Consequences question, sample model-building
   process, 86
Consequential functioning, 66–67
Construction management, attributes case study,
   128–138
Construction projects, 104
Consultants, functions of, 10
Content, in FAST model, 70
Context-dependent insight, 17
Contextual purpose, 185
Continental philosophy, 202 n11
Continuous improvement, 17
Contract negotiation, loop-back modeling, 89–90
Contradictions, 158–159
Cook Food, 15
Cooperative organizational culture, 11
Core technologies, changing, 154–155
Core value, 193
Corrective action, 84, 175
Cost control strategy, 44
Cost drivers, 130
Cost estimate, 56, 106, 109
Cost reduction, 26–27, 29, 32, 60, 85–86, 152
Cost to design, 211 n13
Creative phase, 114
Creative Problem Solving Institute, 33
Creative thinking, 51, 70
Creativity/creativity phase, 13, 70, 109
Credibility, 194
Critical innovation points, 162
Customer(s), generally:
   appeal, 108
   communications, enhancement of, 10
   FAST, 49, 70
   preferences, 20
   requirements, 70
   value, 80

Data classification, 179
Decision(s), *see* Decision-making process
   analysis, 155
   authority, 123

gates, 95
  quality of, 19
  sciences, 17, 202 n8
  technologies, 155
  tree, 155
Decision-making complexity, 124–126
Decision-making process:
  attributes and, 117, 130
  dimensions in, 99
  fuzzy problem technique, 32
Dependency/dependent functions, 60–61, 64–67,
    70–71, 162, 181
Descartes, 203 n13, 206 n6
Descriptive statements, 86
Design, generally:
  configuration, 60
  reviews, 95, 152
Design-to-cost (DTC) targets, 111–114
Desires, 178–179, 186
Detached retinas, case illustration, 159, 161
Dewey, John, 27
Diesel engines, 198
Differentiated products, 51–52
Dimensioning:
  benefits of, 151–152, 189, 224
  business process, 95
  case-specific, 93, 109
  clustering functions, 104, 111–114
  defined, 156
  facility management case study, 96–99
  hard study issues, 94
  operating expenses, budgeting process, 109–111
  organizational effectiveness case study, 101–104
  pipeline case study, 108–109
  pre-event stage, 93–94
  product- and equipment-based models, 104–106
  RACI/RASI, 99, 101, 103–104, 106
  sensitivity matrix, 95–96, 103–104, 106,
    108–111, 114
  soft study issues, 94–95, 114
  staple remover case study, 106–108
  themes, 95
Dissatisfaction, source identification, 151–152
Divergent thinking, 29, 38
Door, basic function, 58
Double-loop learning, 194, 199
Drilled down FAST models, 11
Drilling technology, 157
Drucker, Peter, 181, 183
Duplicate functions, 80
Dynamic relationships, 121, 123

Effectiveness, organizational, 11
Elegant design, 44
Elegant solutions, 61, 70, 89, 162, 187, 199
Emerging technologies, 12, 187
Emerging value methodology, 13
Emotional reactions, 10

Enabling innovation:
  case study, 164–175
  FAST models, 151–153, 162–163
  idea generation, 153–155
  negative functions, 155–159
  outcomes, implications of, 153
  pre-event phase, 163–164
  problem identification, 160–162
Engineering applications, generally, 60. *See
    also specific types of engineering*
English language syntax, 89
Enquiry phase, 85, 194–199
Equipment-based FAST models, 104–106
Erlicher, Harry, 12
Erroneous information, 180
Expansion process:
  duplicates, 79–80
  number of, 78–79
  purpose of, 79–80
Expectations, 52
Expense allocation, 95
Explicit functions, 58
Explicit information, 213 n4
Explicit knowledge, 181, 186, 193, 213 n4
Exploration drilling, loop-back modeling, 90–92
External market customers, 116
External validity, 22
Extrinsic functions, 49, 70

Facility management case study, 96–99
Failure, prediction of, 11
Failure modes and effects analysis (FMEA), 155
Fallon, Carlos, 207 n1
FAST diagrams:
  FAST models distinguished from, 21–23
  validation of function models, 23–24
FAST modeling capability, benefits of, 176
Feedback, importance of, 188
Fifth Discipline, The (Senge), 209 ns16–17, 214 n14
"First to market" products, 47
Flags, 75–76
Focus groups, 118
Focus, significance of, 10, 14, 51, 58, 66, 94, 163
Follet, Mary Parker, 186
Ford, Henry, 195, 215 n16
Framing meeting, 10
Frequency problems, 17
Function, *see specific types of functions*
  defined, 29, 115
  dependencies, 153, 185
  family tree, 39
  identification, 53
  justification, 66, 94
  language, 10
  meaning of, 2–4
  modeling, 15–17, 21–24
  verb-noun technique, 30–31

Function analysis (FA):
  benefits of, 13–14
  defining functions, 48–49
  function identification example, 53
  nouns, 46–47
  random function determination, 53–58
  significance of, 44–45, 58
  syntax, 45–48
  two-word descriptions, 45, 47–48
  types of functions, 48–50
  verbs, 45–46
Function(al) analysis system technique (FAST),
    generally:
  analysis, 7–9, 13, 44–58
  applications, generally, 1–2, 6–7, 9–11, 17–18
  articulation theories, 71–72
  "as is" *vs.* "should be" models, 62–63
  automotive parts case study, 191–193
  build how and test why, 80–83
  building preparations, 80
  building process, product example, 77–80
  defined, 1–2
  development of, 12–13
  function, meaning of, 2–4
  fundamental questions, 18–21
  graphic layout of, 68
  hardware products, 7–9
  how-why questions, variations of, 72–77
  innovation, 6–7, 13–15
  intuition distinguished from, 177–180
  key elements of, 67–71
  logic, 4–6, 10
  loop-back modeling, 89–92
  misconceptions about, 61–62
  model building, 193–194
  notations, 86
  organizational models, 11–12
  practical ingenuity, 17, 159
  problem identification, 83–86
  problem *vs.* opportunity, 21
  process overview, 59–61
  purpose of, 14
  reading, 4–5, 7
  rules, 86–89
  symbols, 86
  syntax, 63–67
Functional enquiry, 194–199
Functional gaps, 66
Functional innovation, 156–157
Functionalism, 18
Functionality, management of:
  components of, 27, 185–186
  shared understanding, 187
  supply chain design, 186–187
  systemic nature of, 20–21
  value management, 187–191
Functional relationships, 125–127

Function cost:
  identification of, 109
  matrix, 56–57, 223
  metrics, 164
  relationship, 63–64, 106–108
Function-speak, 4
Function-weight relationship, 63
Funding approval, 99
Funding stages, 95
Fuzzy problem technique:
  defined, 28, 32
  within hierarchical technique, 38–42
  setting up the problem, 33–38
  significance of, 31–32

Gatekeepers, 11, 109
Gates:
  boundaries determination, 109
  requirements of, 11
General Electric (GE), 12
Generalist knowledge, 182, 188, 213 n8, 214 n10
Generic nouns, 67
Goal(s):
  achievement of, 183–184
  commitment to, 95
  design-to-cost, 114
Goal-setting, importance of, 11, 86, 143, 165–166,
    191
Goldratt, Elyiehu, 179
Graphical language, 71–72
Graves, Clare W., 216 n22
Gullibility, 194

Hard study, dimensioning, 95
Hardware components, 60–61
Hardware product, sample model-building
    illustration, 7–9, 76–80
Haystack Syndrome, The (Goldratt), 179
Heidegger, Martin, 202 n13, 206 n6
Hierarchical diagram, 42
Hierarchical technique, 28, 38–42
Higher-order functions, 80, 85, 144
Higher-order functions, sample model-building
    process, 85
Highest-order function(s), 69
High-level functions, 196
High school renovation, case illustration, 27
High-technology industries, 47
High-value products, 52
Home Bread Machines (HBM), pre-event phase
    illustration, 163–164
Hospital operations, case illustration, 196–197
How questions:
  characterized, 5–6, 106, 117
  function identification, 53
  fuzzy problem techniques, 34–38
  sample model-building, 80–81
How to do problems, 18

How-why dependency, 225
How-why logic, 9–10, 13, 66, 69–70, 121, 136,
   162, 166, 168–169, 178–179, 186
How-why questions:
   purpose of, 18, 60, 78, 86, 88
   reading guidelines, 63–64
   why-how orientation *vs.*, 64–66
How-why relationships, 69–70

Idea generation, 153–155
Ideation International, 159, 212 n3
If-then:
   logic/logic path, 66, 71, 95, 178–179, 186, 196,
      198–199, 225
   rule, 87–88, 136
Ihde, Don, 217 n24
Implicit knowledge, 181, 186, 189, 193, 199
Incentive earned points, 146–149
Incentives, attributes and, 144–145
Independent functions, 66–67, 71, 87–88, 167
Inference, 69, 86
Information quality, 95
Information technology, 101
Ingenuity, 17
Innovation:
   attributes and, 130
   breakthrough, 28
   clustering function, 114
   components of, 6–7, 13–15, 22, 59
   continuous, 17
   coordination model, 23–24
   critical innovation points, 162
   dimensioning process, 96
   enabling process, *see* Enabling innovation
   functional analysis, 45
   influential factors, 94
   knowledge and, 180
   logic path functions, 71
   loop-back modeling, 92
   management of, 13
   problem-solving techniques, 43
   rapid, 115
   situation questionnaire, 159
Input dependencies, 11
Insight(s):
   attributes and, 133
   causality and consequential functioning, 66
   creative, 161
   functional analysis syntax, 46
   implications of, 152
   management, 23
   problem-solving techniques, 28
   project planning, 99
   supply chain improvements, 186
Inspections, 11, 95
Intangible resources, 189–190
Intellectual property, 139
Intended action, 29

Intent explicit design, 20
Intentionality, 167, 186, 195, 199
Intentions, 178–179
Interdependency, 60
Interdisciplinary teams, benefits of, 4, 10, 16–17,
   92, 176
Internal consensus, 22
Internal customers, 116
Internal validity, 22
Interorganization team, functions of, 187
Interpersonal communications, 176
Interrelationships, fuzzy problem techniques, 33
Intrinsic functions, 44, 49–51, 58
Intuition, 64, 177–180
Intuitive knowledge, 178
Invalid data, defined, 179
Invitation to bid (ITB) process, 138–139
I stem, 182

Japanese manufacturing, 49
Jargon, 60
Judgment calls, 70, 211 n8

Knowledge and Its Limits (Williamson), 178
Knowledge base, 13, 20, 132, 159
Knowledge boundaries, 111
Knowledge management:
   benefits of, 23, 212 n4
   new knowledge, discovery of, 180–184
   types of knowledge, 177–181

Lagging effects, 131
Language, communications and, 15
Leadership roles, 103
Lean thinking, 24
Learning cycle, 181
Learning program, 13
Left scope line:
   attribute boundaries, 144
   basic function, relationship with, 81–82
   defined, 69
   innovation points, 162
   sample model-building process, 81–84
Leonard, Dorothy, 182, 185, 214 n10
Levels of abstraction, defined, 54. *See also*
   Abstraction level
Line and staff departments, 183
Logical falsification, 18
Logical relationships, 28
Logic flow, loop-back modeling, 90
Logic path:
   functions, 66, 71, 75–76, 78
   implications of, 4–6, 9–10
   rules, 60–61
Long-term planning, 21

Loop-back modeling:
  characterized, 89–90
  contract negotiations, 90
  exploration drilling, 92
Lowest-order function(s), 69

Major logic path, 71, 75, 88, 117, 156, 162,
  167–168, 225
Malfunctions:
  identification of, 18, 42
  problem-solving process and, 32–33
Management analysis, 94
Management by objectives (MBO), 183–184
Management consultants, 161
Management "silos," 184, 196
Management theories, 123–125, 128, 160, 186
Managerial applications, 60–61
Manufacturing applications, 60, 83
Marketing applications, 60
Market research, 49
Market value, loss of, 52
Matrix organization, 183
Means-end logic, 49–50, 162
Means-end process, 69
Means-end relationship, 32, 39, 60, 64–65, 70, 79,
  88
Mechanical systems, 61
Mechanical use functions, 50
Microanalysis, 10
Microeconomics, 132
Miles, Laurence D., 12–13, 45, 48–49, 54
Milestone events, identification of, 11, 95, 128
Mind, defined, 178
Minor logic path, 71, 75
Mission statement, 86
Monte Carlo simulation, 205 n16
Mousetrap, case illustration:
  functions of, 7–9
  fuzzy problem technique, 34–38
  hiearchical technique, 38–39
Multidisciplinary team:
  benefits of, 4, 10, 42, 60, 62, 177
  shared understanding, 187
Multiple-dimensioned FAST models, 121
Multiple-perspective theory, 183
Myopic management, 206 n7

Nameplate, problem-solving illustration, 30
NASA, 44, 70
Negative functions, 155–159
Neutral function language, 10
New Economy, 214 n10
New knowledge, discovery of, 180–184
New product development, 60
Newton, Isaac, 178
Nike, 177
Nonaka, Ikujiro, 181, 213 n6
Non-value-added functions, 95

No-risk decision, 19
Notations, in model-building process, 86–87
Nouns:
  abstract, 67
  common, 47
  measurable, 46–48
  in problem statement, 85

Objectives, 11. *See also* Goal-setting
Objectivity, 10, 22
Offering, 187, 189–190
Oil drilling, as negative function, 157
Oil filter, basic functions of, 53–54
Oil production, improvement strategies, 187, 196,
  198
Ontology, defined, 204 n15
Operating expenses, budgeting process, 109–111
Opinion-based decisions, 18
Opportunities:
  clarification of, 32, 49
  distinguished from problem, 21
  identification of, 27–28, 85, 89, 163
Organizational boundaries, 139–140
Organizational capability, 188, 196
Organizational culture, 11
Organizational design, 24
Organizational effectiveness case study, 101–104
Organizational FAST models, components of,
  11–12
Organizational functioning, 124–127
Organizational goals, 86
Organizational structure, 27, 96, 186
Organizational success, 189
Organization chart, 11, 184–185
Organization structure, 183–184
Or questions, 89
Or relationships, 73–76
Outlining, policies and procedures, 12
Output dependencies, 11
Outside-the-box thinking, 31, 159, 180
Outsourcing, 191, 196
Ownership, psychological, 32

Paired comparisons, 119–120, 136, 145
Paperwork flow, 95
Paperwork reduction studies, 95
Participative inquiry, 118
Passive verbs, 45, 79
Pencil, case illustration:
  functional analysis, 55–57
  how-why orientation, 64–65
Perceived answers, 85
Perceived value, 10, 52, 108
Perceived workload, 124–125
Performance, generally:
  color coding, 104
  defined, 115
  goals, 11, 191

measurement, 124
metrics, 152, 196
standards, 189
Personnel loading, 95
Persuasion, 22
PERT, 210 n3
Pilot systems, 22
Pipeline case study, 108–109
Polices, development of, 12
Popper, Karl, 180, 202 n9
Porter, Michael, 209 n18
Positive functions, 157–159
Positivism, 204 n13, 205 n15
Post-It notes, model-building process, 80–81, 169
Postmodernism, 203 n11
Practical ingenuity, 17, 159
Pre-event meeting, 10
Pre-event phase:
    attribute identification, 116
    construction management case study, 129–130
    components of, 207 n3
    in dimensioning process, 93–94
    importance of, 28, 163–164
    role in FAST modeling, 143
Preconceived ideas, 67
Preconceived solutions, 129
Prioritization, 10, 61, 166, 194
Problem *vs.* opportunity, 21
Problem formulator, 159
Problem-framing phase, 163, 175, 182
Problem identification:
    fuzzy problem techniques, 33–34
    hiearchical techniques, 39
    importance of, 9, 85–86, 160, 162
    problem set matrix, 160–161
    solution looking for a problem, 161
Problem scoping, 34
Problem set matrix, 160–161
Problem-solution match, 161
Problem-solving approaches, 18
Problem statement, 34, 84–85, 160–161
Problem-solving techniques:
    fuzzy problem, 28, 31–42
    hierarchical, 28, 38–42
    implications of, 42–43
    overview of, 26–28
    verb-noun function, 28–31, 38–42, 48, 60
Procedures, development of, 12
Process flow diagrams, 27
Process flowcharts, 64
Process gates, *see* Gate requirements
Process model, 4, 10, 118
Process phasing, 95
Process vulnerabilities, 11
Produce Tea case illustration, 2–6
Product, generally:
    components, 54–55
    cost, basic function and, 51–52

demand, 195
design, 52
development, 104
improvement, 62–63, 104
Product-based FAST models, 104–106
Production planning, 95
Production schedule, 11
Project completion, 117
Project management, functions of, 183, 187–189
Project plan, development of, 6, 99
Project-sensitive attributes, 121
Proper function, 38
Proposal selection menu, 175
Prototypes, 22, 208 n7

Quadrants, in problem set matrix, 160–161
Quality, generally:
    assessment of, 116
    assurance manuals, 184
    function deployment, 50
    influential factors, 94
    in-process, 186–187
    issues, 87–88
Questions, *see specific types of questions*
    frequently asked, 222–225
    fundamental, 18–21
    sample model-building process, 85–86
Quick-fix solutions, 16–17, 153

RACI matrix, 121
RACI/RASI dimensioning, 99, 101, 103
Radical innovation, 154–155
Random function determination (RFD):
    abstraction levels, 54–55
    characterized, 60, 223, 225
    function and component selection, 55–56
    function cost matrix, 56–57
    implications of, 180
    process simplification, 56, 58
    sample model-building process, 77–78
    selection of, 166–167
Range of acceptance, 118–120
Range of goodness table, 117, 119
Rapid innovation, 216 n20
RASI matrix, *see* RACI/RASI dimensioning
    automotive parts case study, 191–193
    characterized, 121, 183
    modified, 140, 144
Rationalism, 203–204 n13
Reading guidelines, 72, 75, 78–79, 169
Realism, 204–205 n15
Reality, 178–180
Realization, 23
Redefined problem, 161
Reengineering projects, 95
Relationships, identification of, 28. *See also*
    specific types of relationships
Relativism, 202–203 n11

Reorganizated companies, 101, 103, 186
Repetitive functions, 79–80, 89
Request for proposal (RFP), 139
Required data, defined, 179
Required functions, 50
Requirements, in FAST model, 70
Research and development (R&D), 38
Research design and development (RDD), 12, 155
Responsibility, 16–17, 95–99, 140
Revenue drivers, 130
Revenue generation, 60
Reverse engineering methodology, 20
Right scope line:
    attribute boundaries, 144
    defined, 69
    sample model-building process, 83, 84
Risk-based decisions, 155
Risk-reward scenarios, 137
Rival products, 116–117
Root cause, 15, 26–27, 86, 163–164

Scientific management, 186
Scope, 14
    hierarchical technique, 40
    lines, 5–6, 34, 67, 69, 116, 144, 206 n6
Screwdriver, basic functions of, 53–54
Secondary functions:
    basic function rules, 52
    defined, 50
    identification of, 223
    implications of, 47, 58, 60, 70–71
    problem-solving techniques, 39, 42
    random function determination, 55
    sample model-building process, 77–78
Semantics, 80, 89, 143
Senge, Peter, 123, 205 n1, 209 ns16–17, 214 n14
Senior management:
    empowerment of, 23
    functions of, 11, 183–184
    innovation case illustration, 175
    as internal customer, 118
    knowledge management and, 179
    organizational structure, 185
    problem-solving techniques, 29, 32
    shared vision, 193
Sensitivity matrix, 95–96, 103–104, 106,
    108–111, 114, 140
Shared understanding, 187, 191
Shared vision, 193
Short-term issues, 152
Should be approach, 211 n7
Should be model, 62–63, 70, 94, 103, 211 n7, 224
Side effects, 155
Simulation models, 22, 205 n16
Single-disciplined teams, 16
Single-loop learning, 196
Six sigma, 24
Skepticism, 194, 202 n11

Social constructivism, 23, 204 n14
Social responsibility, 32
Soft study, dimensioning, 94–95, 114
Soft VM, 203 n11
Software:
    applications, 60
    procurement process, 140
Solar furnace, hierarchical technique illustration,
    40–41
Solipsism, 202 n11
Specialist knowledge, 182, 188
Specifications, in FAST model, 19–20, 70
Speculation phase, 34–35, 163
Staff manager, functions of, 183
Stage boundaries, identification of, 95
Staple remover case study, 106–108
Starter kit functions, 79–80
Startup companies, 161
Sterling engine, 180
Strategic problem framing process, 28
Strategic problems, 32
Strategic value, 123, 152
Subfunctions, 50
Subjective nouns, 46
Subjectivity, 22
Subproblems, in fuzzy problem technique, 33–34
Subsystems, in clustering, 111, 114
Success factors, 94, 189
Sullivan, Louis, 210 n2
Supply chain:
    impact of, 199
    synergy, 177
Support functions, 6, 71. *See also* Independent
    functions
Symbols, in model-building process, 86–87
Symptom(s):
    fuzzy problem technique, 32–33
    identification of, 15
    masking, 26
    solving, 9–10, 16, 26
Systematic innovation, 17
System dynamics (SD) model, 26, 85–86,
    124–125, 128, 195–196, 205 n1, 209 n17
System of operations, 159
Systems diagrams, 64

Tacit knowledge, 212–213 n4
Tangible resources, 189
Target costs, 95
Task force teams, 186
Team building, 61–62, 84
Teamwork:
    benefits of, 193–194
    brainstorming, innovation case illustration, 169,
    173–174
    communications, 15–16, 60
    consensus, 10, 12, 48, 143, 225
    FAST modeling capability, 176

interdisciplinary teams, 4, 10, 16–17
problem identification, 27
problem-solving techniques, 29
team makeup, 16–17
Technical analysis, 94
Technical communications, 176
Technical direction, resolution of, 12
Technological advances, 163
Technological thinking, 39
Temporary organizations, 186
Test points, 95
Themes, dimensioning process, 94–95
Theory of inventive problem solving (TRIZ), 158–159
Time-dependent functions, 117
Time-sensitive constraints, 118, 152
T knowledge, 182
Top-down management, 183
Transformation, 189
Transmit Torque, 15
Trial-and-error substitution, 196
Trust, importance of, 187
Two-word function descriptions, 45, 47–48, 89

Uncertainty, in problem-solving, 32–33, 163
Underlying causes, 85–86. *See also* Root causes

Validation:
using dimensions, 94
of FAST models, 5, 23–24
logic path function, 76
loop-back modeling, 90
Validity:
how-why questions, 64
logic path and, 64
test of, 22
Value-added functions, 95
Value-adding processes, 186
Value analysis (VA):
defined, 13–14
functional analysis, 45, 48
methodology overview, 129
problem-solving techniques, 32
Value creation, 109, 130, 175, 187–188, 196
Value engineering (VE):
basic function, defined, 50
cost reduction study, 152–153
defined, 14
design-to-cost (DTC), 111–114
dimensioning case studies, 108–109, 111–114
functional analysis, 45
function dependencies, 153
functions of, generally, 22–23, 28–29, 35

innovation case illustration, 164–175
model-building process, 85
problem-solving techniques, 32
software acquisition case study, 139–140
Value equation, 48–49
Value management (VM):
defined, 14
functional analysis, 45
loop-back actions, 89
problem-solving techniques, 32
team, problem identification, 27
Value methodology, development of, 14
Value perception, 116
Value stream, 196
Value studies, purpose of, 108
Value study team, 29
Variance analysis, 23, 205 n16
Vendors, 161
Venture capitalists, 161
Verb-noun syntax, 225
Verb-noun technique:
components of, 28–31, 60
within hierarchical technique, 38–42
two-word function descriptions, 48, 89
Verbs:
active, 45–46, 48, 79, 89
common, 47
passive, 79
Vienna Circle, 204 ns11, 15
Vocabularies, communications and, 15

Wall thermostat, basic functions of, 53–54
What to do problems, 18–19
When, generally:
direction, 71, 109, 156
functions 75
logic, 86–88, 95, 121, 178–179, 186, 196, 225
questions, 5–6, 66
When to do it questions, 18
Where to do it questions, 19
Who questions, 16–17
Why questions:
characterized, 5–6, 64–66
fuzzy problem techniques, 34–38
Word images, 6–7
Workflow disruption, 109
Workshops:
FAST model-building process, 76–80
strategic questioning, 86
World, defined, 178

XYZ-3, innovation case illustration, 164–17

Printed and bound by CPI Group (UK) Ltd, Croydon, CR0 4YY

23/04/2025

14660910-0004